Crawler Tractor Scrapbook Part Two

Richard H Robinson

Published by

Country Life (NZ) Ltd

Rotorua, New Zealand
1999

The author's first introduction to a crawler in 1946.
(see Introduction in Crawler Tractor Scrapbook Pt 1)

Crawler Tractor Scrapbook Part Two

ISBN 0-9597908-4-5

First Published in December 1999

by
Country Life (NZ) Ltd
659 Hamurana Road,
R.D.2, Rotorua,
New Zealand 3221.

©Richard H Robinson

This book is covered by copyright. Apart from any fair dealing for the purpose of private study, research, criticism or review, as permitted under the copyright act, no part may be reproduced by any process, without the permission of the author.

Printed in New Zealand by
Rangiora Print
PO Box 152
Rangiora, New Zealand 8254.

Abbreviations & Conversions

It is normal to use measurements etc which are applicable to the period. However when that period covers nearly 90 years and includes metric and imperial countries things are inclined to end in a muddle. For ease of comparison the tendency has been to use the old imperial measurements.

inches = " = in. = 2.54cm
Feet = ' = ft = .305m
miles = 1.61km
miles/hour x 1.61 = km/h
pounds weight = lbs = .4536kg
ton = 1.016 tonnes
gallons imperial = 4.55litres
E & OE = errors and omissions accepted

Crawler Tractor Scrapbook Pt 2

Contents

	Page
Introduction & Acknowledgements	3.
Beach Tractors	5.
Tracks	13.
Clark	21.
Terratrac	22.
Case	23.
Fiat	33.
Fowler, Marshall & Track Marshall	49
Hofherr-Schrantz-Clayton-Shuttleworth	85.
International Harvester	87.
Loyd	105.
Komatsu	115.
John Deere	131.
Lanz	139.
Vickers	141.
Index	160.

Introduction & Acknowledgements

This is the second part of Crawler Tractor Scrapbook and continues a collection of bits and pieces gathered over a period of years, which I have considered worth preserving. The more I have delved into crawler tractors the more fascinated I have become with the subject and the challenge. Many crawler firms have not been covered previously in a general book and getting information has not been easy is some cases. I appreciate time is money and researching through company achives not very productive. However many people have gone out of their way to help and their efforts are appreciated. In particular I would like to thank Tim Wilkes, of C B Norwood, Palmerston North, for help with Fiat, Nelson Iliev of Motor Holdings, Rotorua for help with Komatsu, Graeme Craw, and Graham Blackley for help with Vickers, the many owners of machines who have brought them out for me to photograph and photographers who have supplied photos. Jim Richardson has continued as general adviser.
Particular thanks go to Reed Publishing, UK for extensive use of material from Farm Implement and Machinery Review and Farm Mechanization. Thanks also to N S Komatsu Australia for permission to use Komatsu material, C B Nowood Ltd for Fiat material, John Deere Australia for John Deere material, Case Corp. and the other companies who have helped.

In a book of this nature it is impossible not to use trade names, however, they are not used officially and are for reference purposes only.
The accuracy of the information I have supplied, particularly in regards to manufacturing dates, is in doubt in many cases. There are so many conflicting statements in many books that it is difficult, to say the least, to decide which is right. I apologise for any inaccuracies.

Crawler Tractor Scrapbook

Farm Mechanization Advisory Service

LONG-LIFE TRACKS
The track pins and bushes on my Fiat 55L have done 3,800 hours and are now worn out. I had them turned after 1,800 hours. Do you consider this a fair working life? My land is heavy but not sandy. The tractor is worked fairly hard but is very well maintained and the driver keeps the tracks on the tight side.

Your track pins and bushes have done well to last 3,800 hours. You were wise to have them turned at 1,800 hours; no doubt this has prolonged their life by some 1,000 hours.

June 1955

CATERPILLAR GOVERNOR
My queries concern the Caterpillar U series D4 and D6. What are the minimum and maximum governed engine speeds of these tractors, and how are the speeds obtained by governor adjustment?

We advise you as follows:—

D4 tractors 6U7050 and up, 7U15662 and up. Low idle speed, 550 r.p.m.; high idle speed, 1735 r.p.m., giving a full load speed of about 1600 r.p.m.

D6 tractors 8U7662 and up, 9U16537 and up. Low idle speed, 550 r.p.m.; high idle speed, 1745 r.p.m., giving a full load speed of about 1600 r.p.m.

D4 and D6 tractors below these serial numbers operate at 550 low idle and 1540 high idle, giving full load speeds of about 1400 r.p.m. in both cases.

The method of checking the engine speed is by removing the cover from the hour meter and checking with a hand tachometer at this point, which is, of course, operating at half engine speed. With the cover removed from the top of the governor the low idle adjusting screw (which is towards the front of the tractor) and the high idle adjusting screw (which is towards the rear of the tractor) are easily accessible for any adjustments that may be needed. After an adjustment has been made to either the low or high idle screw it is advisable to prick-punch the housing near the screw threads and this will hold this particular screw while the second adjustment is being made as, with the cover removed, the adjusting screws may otherwise turn due to engine vibration.

May, 1955

TRACK WEAR
Please tell me the maximum forward movement on the idler of an Allis-Chalmers HD7W to take up track wear, before the pins and bushes are turned.

The maximum permissible movement of the track adjuster before the pins and bushes are turned is $2\frac{1}{8}$ in.

May, 1957

CATERPILLAR FINAL DRIVE
Upon dismantling the final drive of my Caterpillar tractor, I found that the packing between the bearing and the main housing was missing and so allowed oil to escape from the reduction gear housing into the steering clutch housing. Could you please advise me which is the correct type of packing and how to ensure that it stays put?

When you refer to packing we are not sure whether you mean the gasket or oil seal next to the final drive pinion bearing, but in either case the best method of obtaining the correct type is to order from the tractor manufacturer or his dealer, and quote tractor serial number with order.

Regarding your question on how to prevent the packing from moving, the answer is that if the housing and studs are in good condition and assembled in the correct manner, which will be apparent on inspection, voluntary movement of the gasket and oil seal will be impossible.

December 1952

I.H. INJECTORS
I have an International BTD 6 and want to know the timing position of the injectors, and also the injection pressure. In addition, what should I look for when testing the spray?

The C.A.V. spill timing cut-off for the pump is 18° before top dead centre, and the injection pressure is 720-750 p.s.i. The spray from this injector should be fully atomized and even. If it dribbles, is streaky or lopsided, then the injector is faulty. The most probable cause of these faults is an excess of carbon at the nozzle.

May 1959

When speading or leveling the earth, always drive the machine in low gear.

QUERY
I intend to buy a tracklaying tractor and would like some information:—

(1) Among the various tracklayers available at £1,000 or less, new or reconditioned, is there any particular make noted for its track durability, as I gather track maintenance and repair is the chief overhead charge?

(2) Is there any make noticeably more stable than others in 1 in 3 slopes, across as well as up and down, or are tracklayers as a class much of a muchness in this respect?

(3) Finally, in the case of tracklaying tractors fitted with flywheel-assisted splash lubricated engines, what is the effect of steep gradients on such an engine?

ANSWER
We assume that you are interested in machines fitted with the orthodox twin rail track and, therefore, confine our remarks to this type.

So far as new tractors are concerned, there is little difference in track durability. In the case of reconditioned tractors, great care must be taken in inspecting the track mechanism as there are some faults which, though apparently insignificant, can make track maintenance a very costly business. The following observations should help you to detect faults in this respect.

(1) The track shoe grousers should be parallel with their base—they often wear at an angle due to misalignment.

(2) There should be no wear on the face of the bushes. Examine these carefully as they may have been turned, in which case the worn parts would be hidden by the shoe.

(3) Bottom roller wear is usually apparent first on the rear roller. If any of the rollers are worn more to one side than the other, it will denote track misalignment. Another form of misalignment makes itself apparent by allowing the side of the sprocket to foul the track links.

(4) The sprocket teeth should be the same contour at each side.

(5) To test the lateral rigidity of the track frame, turn the tractor as sharply as possible and note if there is any closing of the gap between the radiator base and the inside of the track. If there is, it means that the frame anchorage has moved.

(6) The transmission of a tracklayer is vulnerable. An indication that it is in a shaky condition will be given by the following oil leaks after the tractor has done several hours hard work: oil from behind the sprocket, from beneath the steering clutch housings and from the end of the sprocket axle.

A point to consider when buying a new or second-hand tracklayer is the sales and service organization behind it. The tractor should be of reputable make and there should be a well-equipped dealer in your locality.

So far as suitability on gradients is concerned, there is little to choose between similar makes. As a general rule, of course, the wider machine is more suitable across hillsides.

The effect of operating a flywheel-assisted splash lubricated engine on extremely steep gradients is that the oil tends to flow from the flywheel which, as you know, supplies the engine lubrication pipe. We believe that some manufacturers advise that an extra quantity of oil should be put into the sump to offset this.

We suggest that you raise this point with the manufacturer, who will be pleased to give you specific recommendations.

March, 1951

Beach Tractors

I mentioned to some of the chaps that I was short of colour photographs of crawler tractors.

"Go out to Ngawi ." came the instant reply from two.

"Where on earth is Ngawi?" I asked. It is pronounced Nar wee and spelt Ngawi or Ngawhi.

"Down the bottom of the North Island along the beach from Lake Ferry."

Sounded like a good trip to me and was in the area I had other machines to photograph so I took three days off.

I left camp at Lake Ferry in light fog but by the time I arrived at Ngawi it had lifted and the sun came over the hills by 9.30 giving me a beautiful morning.

Crawlers of all descriptions were parked up on the beach ready to launch fishing boats. I counted 38 in various stages of mobility. It was obviously the semi retirement beach resort for crawler tractors and also the last resort.

There was only one wheel tractor, a Massey Harris 81 and not many places where it was hard enough to use it. The crawlers were ideal for the gravel conditions, the blades were put to good use each time a boat was launched.

Ngawi

Crawler Tractor Scrapbook Pt 2

Captain Morgan David Brown 30's 41 brake horse power is sufficient to launch this fishing boat into Cook Strait.

I am pretty sure "Kermit" is a Track Marshall 55. It is powered by a 55hp Perkins 4 cylinder diesel engine and they were made between 1959 and 1970.

This rather sleepy looking crawler is an Allis Chalmers HD5. They were made from 1946 to 1955 and had a 30dbhp engine.

Beach Tractors

David Brown crawlers are either plentiful or cheap to buy because there are quite a few on the Ngawi beach. This David Brown 30 has had its day and is now only suitable for parts.

Above and beside both these David Brown Trackmaster 30s are still in use but one wonders for how much longer. It would appear that it is not necessary to have tight tracks.

Considering the work that some of the smaller crawlers are expected to do on the beach this Caterpillar D6 looks like an overkill. The D6 tractors were made from 1941 to 1959. They had a Caterpillar six cylinder diesel engine with a 4½ inch bore and 5½inch stroke giving a rated dbhp of 59.33. It had five forward speeds from 1.7 to 6.6mph.

Above: Someone has felt inventive and made a substantial push bar to fit on this Case Terratrac 601 loader, I think they were made from 1957 to 1962. Next to it is an Oliver Cletrac and further on probably a David Brown.

Left: As the boats get bigger so do the crawlers. Obviously there has been some problem with this Allis Chalmers HD16, hopefully only the battery. The HD16 were made between 1955 and 1961 and had 105dbhp.

Beach Tractors

Above: A Track Marshall waits for the next tide. It would appear that number eight wire has gone out of fashion, baling twine holds the air cleaner in place.

Left: A David Brown 30TD requires blocks behind its tracks suggesting that the brakes are not very efficient.

Below: Is this a Caterpillar D4 or a Caterpilar D6? Which ever it is there is plenty of power for that sized boat.

Crawler Tractor Scrapbook II — Beach Tractors

The Allis Chalmers HD5 crawlers are popular as boat tractors. Here are two more. The HD5 were powered by a GMC two cylinder diesel engine with 4¼ and 5 inch bore and stroke giving 48 brake hp. They had five forward speeds from 1.5 to 5.5mph.

Just to show how it is done an early Komatsu D50 launches a fishing boat, or maybe he was just going out to clear his crayfish pots. When they are ready to come back in they just ring up on the cell phone for the crawler.

Crawler Tractor Scrapbook Pt 2

Beach Tractors

A Caterpillar D6B waits for its owner to return from the sea. The D6B was made in USA, Australia and Japan from 1959 to 1968 then in France from 1977 to 1979. It had a Caterpillar D333 engine giving 90/93 engine hp. It weighted 24,300lbs.

Middle: An Allis Chalmers HD11EP and a Caterpillar D6. The HD11EP was made from 1961 to 1971 and had a 195dbhp engine.

Bottom: A Caterpillar D6 waits.

Crawler Tractor Scrapbook Pt 2 — Beach Tractors

The Royal National Lifeboat Institution (RNLI) in the UK commissioned several Fowler Challenger IIIs to be built for launching lifeboats into the sea. There were other makes of crawlers adapted at various times such as Case.

The Challenger III was built in 1953 and several modifications followed. The most important part was the waterproofing of the engine. VYT 878 was built in the later part of 1954 or 1955.

Photo: Richard Trevarthen's collection.

When parking for any length of time on a hill or slope, be sure to lock the brakes on before getting off the Bulldozer.

Everybody who has the least sensibility or imagination derives a certain pleasure from pictures.—MACAULAY.

IT was in our last issue (page 1062) that we mentioned the ample illustrations employed by the Marshall Organisation in their house magazine *Mettle*, and one series in particular in a recent number tickled our sense of humour, for it tells "the saga of a 'Challenger'—and its crew—which went to sea!" It seems that one of the requirements of the Royal National Lifeboat Institution is a tractor that will launch a boat from difficult beaches and irrespective of tidal conditions, and so Messrs. John Fowler & Co. (Leeds), Ltd., were invited to prepare a waterproofed version of the "Challenger 3" that can be driven into the sea and, if necessary, left submerged without any qualms, to be retrieved later after the tide has receded. A special "Challenger" was, therefore, taken to Aberystwyth for testing, and with a crew that included Mr. Hopkins, their chief engineer, and Capt. Michelmore, of the R.N.L.I., it was driven out for an immersion test with everyone looking quite comfortable and suitably arrayed in waterproofs, gumboots, etc.

* * * *

Upon the word,
Accoutred as I was, I plunged in.—SHAKESPEARE.

UNFORTUNATELY, the chosen stretch of beach shelved much more sharply than the local experts had warned, so that the tractor suddenly disappeared beneath the waves and "its crew lost no time in obeying Capt. Michelmore's instructions to 'abandon ship'." The last photograph shows the gallant band swimming ashore, still wearing trilbies and caps, but looking as determined as if it were a cross-Channel attempt, while to one side is a periscope-like object that is evidently the tractor exhaust pipe.

* * * *

Down, down beneath the deep.—H. F. LYTE.

HOWEVER, "all's well that ends well," and, not only were there no casualties, but observers were sufficiently impressed by the tractor's performance to order five such machines. Some slight modifications to the controls of the waterproofing gear have now been made, including the addition of a light to enable the driver to see what he is doing (if not where he is going!), even though his whole cockpit is below the surface. We can imagine orders coming in from those who wish to practise that latest of sea sports, underwater fishing, in a leisurely fashion, but we are quite sure that, whatever reason Mr. Hopkins may have to make a similar trip, he will do it in bathing trunks and only after making sure of a warm day.

TRACKS

Dec. 1, 1951 — FARM IMPLEMENT AND MACHINERY REVIEW

The Bryden "Tracpak"

Gives Full Crawler Track Conversion for Wheel Tractors

An entirely new and simple form of converting standard wheel tractors to crawlers has been devised by Messrs. George Bryden Engineering, Ltd., York Road, Seacroft, Leeds. Primarily it has been arranged for the "Ferguson" as known in this country, but it is also applicable to other machines, such as the American "Ferguson," the "Ford 8N" and virtually any wheel machine. Even now, the makers are designing a larger unit for a high-powered tractor.

The inspiration behind the new development is Mr. George Bryden, M.I.A.A., M.I.B.A.E., who will be well remembered in the trade for his association up to a little while ago with the Marshall Organisation; he had much to do with the development of the "Fowler-Challenger" range of tractors. Now he is concentrating his efforts, in conjunction with those of his eldest son, in designing what is known as the "Tracpak," which affords ready means of obtaining crawler operation from an otherwise standard type of 4-wheeler. Choice of the "Ferguson" for the launching of the idea was adopted for obvious reasons, such as its existing popular sale throughout the world.

Animating the move is the belief that full track machines *per se* are an unnecessary capital expense on any farm where a light tractor offers sufficient drawbar power in the ordinary sense, and where the major amount of work can be satisfied by a wheel machine. But it is appreciated that there are instances and also occasions and soil circumstances when the availability of a track would be of untold benefit, even if only for temporary duty, and that is why in this easily convertible full track unit care has been taken to see that nothing is done to impair the versatility or allround utility of the original prime-mover, or to introduce anything requiring skilful manipulation or intricate parts; consequently, there is concentration on standard BSF bolts, nuts and pins.

Furthermore, every implement used by the "Ferguson" tractor in the ordinary sense when on wheels can be applied with similar efficiency with the "Tracpak," while spring type on-and-off street plates of a boltless design are available for the track, so that it can be converted to go on the road in 10min. (street plates can be put on and off both tracks in 15min.). Again, the "Tracpak" can be placed on and off a tractor by two men in 3½ hours (from wheels to tracks 2 hours; tracks to wheels 1½ hours), whereupon it is again in its wheel form.

Neither alloy nor carbon steel is used and approximately only 2,000lb. of M.S. per "Tracpak" is required, consisting of "T's," tube, flats, rounds and squares, all of them standard stock rollings, so that obviously here is a development that should be of inestimable value in saving steel at this critical production and supply stage. Not only is the design what might justifiably be described as a "steel saver," but it is also of value from the export angle, in that the demand from overseas is already considerable.

TRACKS around WHEELS—

—with all Wheels Driven

THE tractor shown here is an old I.H. TD-6 fitted with prototype tracks designed and provisionally patented by Mr. Peter H. Matthews of Adelaide, South Australia.

After years of land clearing with tracklayers in his own country Mr. Matthews sold out and came to England convinced that he could produce a track which would combine the tractive efficiency of the orthodox track and the maintenance economy of the pneumatic tyre.

In conjunction with C. Norris and Son, Ltd., Wellingborough Road, Rushden, Northants, who provided the engineering facilities, Mr. Matthews has converted the TD-6 which *Farm Mechanization* photographed while it was being field tested.

Chain Drives

An important point is that *all the wheels are driven*, so the track is not subjected to the same type of stress as the driven pin-and-bush track. The present drive is by chain, from the rear, between the wheels, but this arrangement is to be superseded by chains in oil-bath housings situated between the inside wheels and the chassis. Field tests have shown that the track will not run off in rough terrain. For dry conditions, the tracks can be removed and the tractor utilized as a *six-twin-wheel-driven unit*.

Mr. Matthews and Norris's would be pleased to hear from a manufacturer interested in furthering the development and production of this type of track assembly. Another type of track, with grouser plates and girder action, is also being developed.

With plough and dynamometer, Mr. Matthews tests the performance of his track which proved capable of stalling the engine of the old TD-6.

Farm Mechanization April 1961

Agricultural Machinery Patents

Improvements in Tractors

Maurice Victor Poncet, 37, Rue Victor Hugo, Villefranch-sur-Saone (Rhone), France. (636,228.)

This specification covers a method of adjusting the width of tracks on a crawler tractor without the use of any kind of tool and even without stopping the tractor. The adjustment is made by a number of hollow transverse shafts which are screw-threaded internally and externally with equal but opposed pitch, each co-operating with a screw-threaded sleeve and rod attached to the track frames.

British Farm Mechanization February 1950

AN EARLY TRACKLAYER

Sir,

In connection with the Golden Jubilee of the Caterpillar tractor, it is perhaps of interest to recall that 184 years ago (in 1770) Richard Lovell Edgeworth patented a method of " making portable railways to wheeled carriages, so that several pieces of wood are connected to the carriage, which it moves in regular succession in such manner that a sufficient length of railway is constantly at rest for the wheels to roll upon." One hundred and thirty-four years were to pass before the idea was really used.

Yours, etc.,
EWEN M'EWEN.
*Department of Agricultural Engineering,
University of Durham,
King's College,
Newcastle upon Tyne.*

Farm Mechanization May 1954

Farm Mechanization April 1961

Right: Another type of track better suited to haulage in snow than for farm cultivations: it was seen fitted on a Soviet tracklayer.

Spectacular Tracklayer

The most spectacular performance on slopes and over ditches was made by the Rolba Bombardier Muskeg tracklayer made by Bombardier Snowmobile, Ltd., Valcourt, Quebec, Canada, and distributed in Britain by W. E. Weisflog, Rolba and Bombardier Equipment, 88-92 Rochester Row, Victoria, London, S.W.1.

Designed primarily for snow, slopes and soft land this tractor has tracks made of rubber and rayon reinforced

with steel cables and fitted with steel grousers. The carrier wheels have 4.50 x 16 pneumatic tyres and the tracks are driven at the front by fabric-reinforced rubber sprockets.

Although this tractor weighs 4,700 lb. it is said to have a ground pressure of less than 1 lb. per sq. in. The payload capacity is given as 2,500 lb. A Chrysler 115-h.p. engine is used, but a Perkins Six 305 diesel will also be made available. Steering is through a differential and the maximum speed is 25 m.p.h. The price is expected to be about £2,600.

Farm Mechanization June 1960

FARM MECHANIZATION   October, 1955

Retreading Tracks—Welding Time Halved

From material prepared by S. M. Algar, R.I.B. Technical Officer (Welding)

THE following methods of retreading track grousers, when using Onions track strip, are advised by the Rural Industries Bureau. The strip is available from Jack Olding and Co., Ltd., Hatfield, Herts (the world distributors).

	Size No.	Dimensions	Cut lengths
For tractors up to 50 h.p.	1	1 in. x ½ in.	12, 13, 14, 15, 16, 18, 20, 22, 24, 26 in.
For tractors over 50 h.p.	2	1¼ in. x ⅝ in.	As above
For badly worn grousers	3	1½ in. x ¹¹⁄₁₆ in.	Cut to customer's requirements

By eliminating the need for hard-surfacing electrodes (as required when ordinary mild steel bar is used) this high carbon steel track strip halves actual welding time.

Method 1. When dealing with detached grouser plates a simple jig should be devised to hold the plates vertically on edge, and positioning blocks provided to maintain the track strip in position during tacking. The track strip should be tilted slightly below the horizontal so that the contraction of the weld pulls it up into the horizontal position. Tack the track strip at each end and then complete welding from one side, afterwards turning over and completing welding on the other side.

Use as large an electrode gauge as possible, preferably a 4 s.w.g. electrode at 200/230 amps. to British Electrode Classification E 217. If the grouser plates to be retreaded are worn curved it will be necessary, first of all, to bring them to a straight edge with a cutting torch.

Method 2. If the grouser plates are bolted to the track chain, hoist the chain vertically by its middle link, with a block and tackle. All welds are now in the downhand position and, working on a pair of steps, the track strip can be both tacked and welded one side. To complete the job, re-hoist the chain by each end in turn to bring the other side of each weld into the downhand position.

Method 3. If the grouser plates are attached to the track chain and the chain is on the tractor, jack up the tractor on to suitable blocks with both tracks clear of the ground. Loosen or remove sparking plugs or Diesel injectors to eliminate compression. Engage bottom gear to enable the chains to be rotated via the starting handle and transmission.

In this way four grouser plates, one at each end of both chains, are always in a flat downhand position for welding. Attach an earth clamp to the tractor side frame to prevent current running through internal bearings and possibly causing damage. Weld the track strip to each of four grouser plates in the downhand position. Rotate the chain by the starting handle until a further four plates come into the same position.

Eventually the two plates which were first welded on their top side only, at the front of the tractor, will be presented the other way up at the other end of the tractor, and vice versa for the two plates first welded at the rear end of the tractor. Welding can then be completed.

Three Days' Work Saved

Methods 2 and 3 can eliminate as much as three days' work for two men in unbolting and rebolting grouser plates: high tensile bolts and nuts are saved, as some usually have to be cut away owing to excessive corrosion or thread stripping; all tendency to distort on the part of the grouser plate is prevented by its being securely bolted to the chain. Subsequent assembly difficulties are also avoided.

Crawler Tractor Scrapbook Pt 2 — Tracks

Farm Implement and Machinery Review.—Feb. 1, 1958.

Italy Produces an Unusual Crawler Track

Details of a new type of track for farm tractors are given by Signor Leo Pardi, of the Pirelli company of Milan, in a special article in *Macchine & Motori*, of Bologna. This is a track that runs on rubber rollers and, indeed, its main components are rubber.

It is said that the conventional pattern metal track has the big disadvantage that it cannot travel freely and effectively on the highway or upon any rigid hard surface. Such travel either damages the road surface or imparts too much vibration and shaking to the tractor itself. These disadvantages and discomfort arise through the too rigid and inflexible contact of the track elements with a hard unyielding surface, with the result that there is a scraping and slipping effect rather than positive grip.

To overcome these and other drawbacks, Count Bonmartini has devised and the Pirelli company have constructed this new type of track, which runs on rollers, whose main element is rubber housed in an original type of holder or base in such a way that the track has free movement to follow the exact contour of any surface, be it rigid or otherwise. The fundamental characteristic of the track, which becomes a permanent feature of the tractor and is in no sense an alternative or subsidiary fitting, is that the articulated links, which make up its continuous formation, are furnished at the point of ground contact with small rubber rollers turning on a longitudinal axis, that is to say, one in the direction of tractor travel.

Thus it is not the link unit itself that bears directly on the ground but the corresponding rollers it contains, each one of which turns round on its own axle every time the appropriate link member is transversely deposited on the ground. The combined movement of the track proper around its own track rollers, and that of the small rubber rollers within their own axis, it is declared, eliminates entirely any tendency to scrubbing on the surface, since all transverse friction is avoided. A big factor in this direction is that, instead of there being the usual inflexible link units bearing on the ground, there are the more adaptable rubber rollers, which conform quite faithfully to the road or other contour and make more complete contact, even in the case of very

A "Fiat" fitted with the New Track

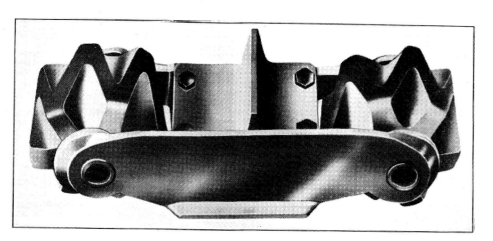

A Close-up View of the Link

Farm Implement and Machinery Review.—Feb. 1, 1958.

minor deflections, than is possible with any other form of track.

Comparative tests between tracks of this design and those of the conventional form have been made and have given excellent results. They have established that, on the highway or similar hard and unyielding surface, there is greater tractive effort, less slip and higher drawbar pull, while, as will be realised, there is the further big recommendation of no damage to the road surface. In farm work, as in running over a newly-cut meadow or working on a stiff short stubble, the tractive effort and the maximum drawbar pull have been substantially increased. In short, it has been established that the new track gives a better pull/weight ratio than was previously obtainable with the standard metal track fitting.

The conclusions are that such a track can reproduce all the effectiveness of the metal one in field work and that in road transport it is far superior, for it can do transport duties prohibited to the metal form and do them, moreover, at a speed comparable with wheel tractors, which is beyond the range of the metal tracked tractor.

In view of this greater mobility, it is but a short step to the further and more intensive exploitation of the idea, for the faster rate of road travel opens up tremendous opportunities for the farmer in speeding up operations, especially when he has to move equipment, if not from farm to farm, at least from one centre to another and traverse roads in so doing. Thus it is discovered that, in the half-track form, the new track, in conjuction with pneumatic tyred wheels on the rear, forms something like an ideal transport unit for combines that have to travel from farm to farm and also have often to work in sticky conditions, such as ricefields, which are unsuitable for wheel machines.

Trials have been made with a Belgian-made combine so fitted, and the results have been described as highly gratifying. Altogether, the track is regarded as opening up new avenues in agricultural engineering, for, apart from making the track tractor more challenging to the wheel one in certain spheres, it releases the track machine from many onerous restrictions, especially on the roads.

In explanation of the schematic drawing it is stated that the traditional metal track is replaced by a series of plates (1), which are articulated and pivotally hinged (2) to form links. Each plate has two projections (3) which support the axle (4) of the rubber rollers (5) and each plate is fitted with two rubber rollers. The forward displacement of the machine is governed by the movement of the track around its own driving wheels (6) while any lateral movement, as in turning, is controlled by the rotary movement of the rubber rollers (5) around their own axles (4). That is to say, the vehicle is free to travel or turn in all senses without friction and in the smoothest possible manner. Each plate is formed at the end with the appropriate gripping or linking rib of the customary type (7), which also increases the tractive effort when working in field conditions, while the centre rib on the support plate (8) acts in its longitudinal position as a rudder in keeping the track to a straight path on sidling land and other hilly conditions. Naturally, the rubber rollers project more from the track than any metal parts so that road contact is exclusively made by them.

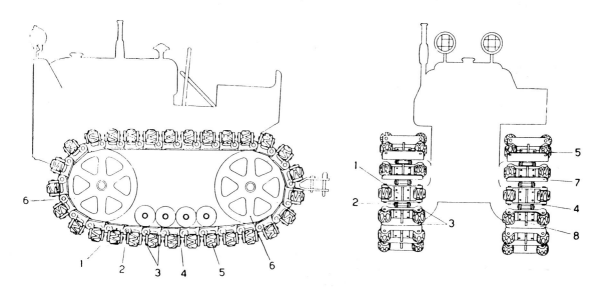

For Key see Text of Article

Farm Implement and Machinery Review.—March 1, 1960.

Two Tractors— in One

Crawler for Hillsides and
Heavy Duty;
Wheels on the Flat and
the Road

Wheel and Track Arrangement Side by Side

ALMOST from the time tractors were first devised there has been controversy as to whether the wheel or the chain-track form is the more suitable allround or all-purpose machine. Many attempts have been made to secure the best of both worlds by trying to make an easily convertible type that could be used at will in either track or wheel form, according to immediate requirements, a development that has become all the more urgent within the last decade or so in view of the increasing amount of road transport work now required by the average farmer from the general-purpose farm tractor.

Years ago, efforts were made to solve the problem of maximum adhesion in field work and alternative complete mobility on the roads by introducing all manner of special types of wheels. Some of these had auxiliary field fittings that were removable when road work had to be undertaken after field operations, so as to give the essential flat smooth bearing surface demanded by road surveyors. But, a common experience, many of these auxiliary fittings got lost in the changeover from field to road duty and *vice-versa*, and the overall results were not satisfactory.

It is true that perfection of this idea has led to the introduction latterly of much more satisfactory forms of alternative adhesion or traction fittings for rubber-tyred tractors in tillage and similar exacting work, and that, as these are retractable after field work when the tractor has to go on to the road, they solve the problem in one way, as do, of course, cage wheels and other extension forms when abnormal soil conditions have to be negotiated at given times of the year and in specific operations.

All this, nevertheless, merely imparts particular interest to a new Italian idea whereby the same tractor can be used either as a crawler or, by a simple changeover, in the wheel form, with the track arrangement advised for heavy and siding land, and the wheel one for work on the flat. The dual-purpose machine is made by Costruzioni Meccaniche Gallamini Vincenzo, via del Milliario, 40, Bologna, Italy, and is said to have 10 speeds, but only eight of these are forward and the other two in reverse.

The innovation, the makers state categorically, consists not in the possible transformation from a track machine to a wheel one — that is not claimed as anything entirely new in itself — but in the fact that, when the machine is operating as a wheel type, the entire chain-track mechanism is completely disengaged from the transmission and is static, so that no power is wasted in operating parts not functioning at a particular moment.

This means that the chain-track mechanism is completely isolated, with the consequence that needless wear is not imposed upon it when it is in fact "idling" or out of engagement. Furthermore, the tracks do not come in contact with the road surface when road transport work has to be done, thus again saving unnecessary wear. The designers make a great point of the fact that, in these circumstances, the crawler track transmission is completely at rest.

Conversion is easily and quickly done, even by unskilled labour, for both mounting the wheel equipment to the basic tractor and removing it is within the capacity of two men. In mounting, the rear wheels and fittings are put on first and then the front ones, but in the removal this process is reversed. In each case, all that is required are the normal tools supplied with the tractor.

It has been shown that the time required is as follows:—

Mounting the rear wheels—2hr. 35min.
„ „ front wheels—1hr. 10min.

Total 3hr. 45min.

Dismounting the front wheels— 50min.
„ „ rear wheels—2hr. 30min.

Total 3hr. 20min.

Stress is laid on the fact that quite clearly the time required for the two jobs is very short, especially for unskilled labour, since the transformation is done quite simply and easily. That is why the makers describe their machine as "a sensational surprise for 1960, giving farmers two tractors in one." In each case, the power-unit, chassis, etc., are common to both forms, with the engine a 4-cycle 2-cylinder diesel, having forced air cooling and giving approximately 25 h.p. at a speed of 2,000 revs. per min.

There is gear pump lubrication, an oil-bath air-filter, Bosch injection equipment, and a 12-volt dynamo battery electric starting system. A dry single-plate clutch and steering discs (working in conjunc-

Crawler Tractor Scrapbook Pt 2

tion with the brakes which act on brake drums), are other standard fittings, and the total weight in working order is 1,400 kilogrammes (3,080lb.). Rear power take-off pulley equipment, a hydraulic-lift, etc., can be supplied as "extras."

The most important characteristic, of course, is the convertibility, and this is facilitated by the fact that basically the front part of the machine follows faithfully standard wheel tractor design; thus it can be easily modified.

For example, as is shown in our drawing, special front wheel steering mechanism, instead of the general track steering controls, can be simply fitted, with everything answering to the driving wheel conveniently placed for the tractor driver in the same position he occupies when tracks are used.

The way in which the machine differs from the standard tractor is that the entire front wheel arrangement can be withdrawn simply by releasing a locking plate or fitting. The rear section consists chiefly of a hub divided into three parts. There is the hub itself, which is keyed on to the shaft extending from the final reduction of the tractor; a flanged grip ring or collar on to which is fixed the driving wheel of the track mechanism, and this ring is co-axial with and rotates on the rollers outside the hub itself; a further flanged ring is placed outside the hub laterally or offset for the fitting of the pneumatic wheels. All these parts are quickly placed in position or withdrawn, simply by fixing or releasing eight bolts.

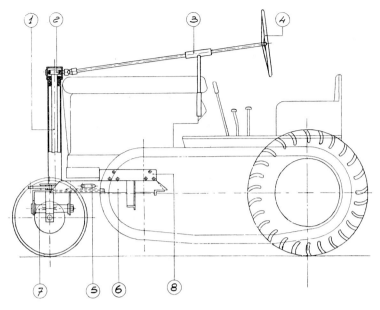

(1) Control Shaft to Steering Wheels; (2) Enclosed Steering Worm Wheel; (3) Steering Rod Support; (4) Steering Wheel; (5) Locking Plate for Front Wheel Brackets; (6) Supporting Member for Front Wheel Fittings; (7) Centre Axle; (8) Chassis Fixing for Front Wheel Supporting Member.

Italian Pneumatic Tractor Track

THE possibilities of a pneumatic track for light tractors were demonstrated recently at Aosta, Italy, during the first international conference on *The Mechanics of Soil-Vehicle Systems*. This pneumatic tubular track has been developed by Signor G. Bonmartini, primarily for aircraft landing gear; it is claimed, however, that it can be usefully applied to tractors, and has been fitted, as illustrated, to a "Lombardini Castoro" light tractor for experimental use.

The track is actually an inflated tubular tyre, the casing being made up of two nylon or rayon plies fastened to two steel bead cables, one on each side of the tube. It is designed for use with bogie wheels shaped in such a way that the bead cables run very close to the wheel rim. The inner surface of the track is lubricated with castor oil by a small pump driven off one of the wheels.

The landing gear of a "Piper" light aircraft, which is used for crop spraying, has been fitted with the "Bonmartini" track. Reporting on the demonstration, R. M. Ogorkiewiz (*The Engineer, July 21st*) says that the machine "successfully crossed small ditches which would have been disastrous for aircraft with the normal single-wheel landing gear."

As applied to the "Lombardini Castoro" tractor, the track drive relies on friction between bars connecting the flanges of the rear driving wheels. Although this imposes some limitation on traction, it is believed that the pneumatic track has considerable agricultural possibility, especially where minimum disturbance of the soil is important. It is suggested that the track might also be used, under certain conditions, for trailers.

Another, more orthodox track designed by Signor Bonmartini was also demonstrated, on a Fiat tractor and trailer, at Aosta. This consists of pin-jointed links, each of which has two rubber-faced rollers free to rotate about pins set in the links at a small angle to the longitudinal axis of the tractor.

The claim made for this design is that resistance to lateral track movement is greatly reduced, and a smaller moment is needed to turn the tractor, on hard surfaces at any rate. But on soft ground, where considerable track sinkage occurs, the rubber-faced rollers lose their advantage, and the track could not be used in vehicles with a greater length to width ratio because they would be as unsteerable in soft ground as similar vehicles with conventional tracks.

Tractor Application of the "Bonmartini" Track

Crawler Tractor Scrapbook Pt 2

Farm Mechanization January 1953

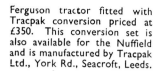
Ferguson tractor fitted with Tracpak conversion priced at £350. This conversion set is also available for the Nuffield and is manufactured by Tracpak Ltd., York Rd., Seacroft, Leeds.

WARCO ALWATRACS

THE MODEL KM

The largest of the WARCO family of Alwatracs. Attached to the 15-30 International Tractor. Has 16" pressed steel semi-flat treads to which lugs can be bolted when required. A pulling unit with power to burn.

THE Models KM and HM Alwatracs can easily be attached to your International Tractor, or the outfit may be bought complete with tractor installed.

These tracks are supplied with tread plates giving sufficient bearing surface that they will not mark the softest asphalt and yet will give ample traction in the most adverse soil conditions. For extreme conditions they can be equipped with our standard lugs or with extra-width treads.

There is a separate clutch in each track, making it possible to turn the tractor under all conditions with a capacity load. It will turn in its own length. The clutches are simple, powerful, and effective, and the control is automatic.

On the 10-20 the bearing pressure per square inch of area is 4.1 lbs.; on the 15-30 it is 5 lbs.

The tractor equipped with these tracks will pull at least 50 per cent more than the standard wheeled tractor, and in many cases will be able to operate where the ordinary wheeled machine could not be used.

W. A. RIDDELL COMPANY
Bucyrus, Ohio

THE MODEL HM

Built for the 10-20 International Tractor. Has 12" semi-flat treads with provision for lugs. The Model KM'S little brother —and not so little, at that!

1928

Clark

Wheel tractors and even crawler tractors have been transported by aeroplane into remote areas but the Clark crawler is the only known one specifically designed for that purpose. It was built in 1942 for the American Army. It was designed to be carried by glider, or whatever, into war zones.

The Clark CA-1 machine weighed 4,196lbs which compares with the Oliver OC3's weight of 4,000lbs. It had a Waukesha FC, four cylinder, petrol engine which produced 28hp at 1,900rpm with a drawbar pull of 4,000lbs again almost identical with the OC3's maximum pull of 3940lbs. It had four forward speeds and one reverse.

Unfortunately Clark Equipment Co. suffered a factory fire early in the contract period and the government handed the contract over to the American Tractor Co. After the war ATC must have seen the possibilities of a small crawler and in 1950 they introduced the Terratrac range of crawlers with the GT25. ATC was taken over by Case in 1957 so the Clark crawler was the fore runner of the Case range of crawlers.

Photos- Jim Richardson

This Clark CA-1 crawler tractor was photographed at The "Gathering of the Orange West" at Marshalltown, Iowa, USA, in 1998. As it was designed for the army it can be assumed that the original colour was army green. It was a 1944 model and was owned by Vernon Waterman.

Photographed at the Henty Rally, NSW, Australia in March 1999 this Clark crawler was owned by Neil Puls. "This Clark was used in Papua New Guinea. While most were buried or dumped in the sea this one found its way to a station on the Queensland border and was used for land clearing.
The boxes on the side were filled with sand as ballast."

Crawler Tractor Scrapbook Pt 2

TERRATRAC

The American Tractor Corporation, ATC, started marketing crawler tractors in 1950 under the name of Terratrac. The first model was the GT-25, a small agricultural machine and was followed by the GT-28, GT-30, GT-32 and GT-34. The GT-25 was listed as a 2 plow, 14 inch moleboard tractor and the GT30 as a 3 plow tractor. The GT-30 had a Continental, four cylinder engine which produced 20dbhp and 26.8 belt hp. It had three forward speeds of 1.78, 2.81 and 4.61mph.

The Terratrac DT-34 was introduced in 1953 and advertised as a three furrow, 14inch plow tractor. It weighed 4,750lbs. The engine was a Continental GD-157 with four cylinders, rated at 1850rpm giving 34hp. It was superceded in 1955.

In 1956 there were eight different models available. The Terratrac 200, 300, 400, 400D, 500, 500D, 600, 600D.

In 1957 Case bought American Tractor Corporation. At that time there were five models marketed ranging from 34 to 62 horsepower and these were continued until 1959.

Terratrac G300, serial #150 supported a business sign when photographed at Mapua, Nelson, NZ in 1999. It had a Continental F-140, 4 cylinder, petrol engine giving 30hp. It was made in 1955.

This Terratrac 300 is serial no. 001. It belonged to Bill Higgs, Nelson, NZ when photographed in 1999. They were made from 1955 to 1957.

Crawler Tractor Scrapbook Pt 2 — Terratrac

Top : The Terratrac 300 weighed 4830lbs and had a Continental, F-162, four cylinder engine rated at 1850rpm. It had three forward speeds of 1.74, 2.75 and 4.52 mph and reverse 2.01mph. It was made between 1955 and 1957.
This machine belongs to Jim Richardson and was photographed in 1988.

Centre and below: This Terra trac 400 was at the Gore New Zealand, rally in 1989. It had a Continental F-162, four cylinder engine with a 162 cubic inch

Case

In 1842 Jeremy Increase Case, at the age of twenty four, started tinkering with threshing machines. That was the beginning of Case Corporation.

Case did not manufacture crawler tractors, as such, until they bought the American Tractor Company in 1957. Other manufacturers track laying conversion units were used on Case wheel tractors. The Case CI was used as a base for conversions to the likes of Trackson tracks from about 1929. The model LI was used as well. The English Roadless used also Case base units successfully.

The Case CI on Trackson tracks pictured here has been adapted from steering wheel control to hand levers.
The engine was a Case, I head, four cylinder engine with a bore and stroke of 3 7/8 x 5½ inches rated at 1,100rpm. Maximum brake hp was 29.81. This machine is part of the Graeme Craw collection, Northland, New Zealand.
Photographs from Graeme Craw's collection

Right: Another Case CI this time on Roadless Tracks and with a steering wheel.
Photographed at an English rally by Richard Trevarthen.

Crawler Tractor Scrapbook Pt 2 — Case

The Case 310 was made between 1957 and 1963 but the 310 G carried on to 1970. It had a Case, four cylinder, engine with bore and stroke of 3 3/8 x 4 1/8 inches giving 33.32 maximum PTOhp and 26.75dbhp maximum. It has three forward speeds of 1.74, 2.75 and 4.52mph.
This machine was at the south coast of the North Island, New Zealand in 1998.

The Case 450 was built between 1965 and 1979 and was followed by the 450B and 450C. It was powered by a Case G-188D diesel engine giving 51hp. Weight was 10,795lbs.
These photos were taken by the owner Bill Higgs, Nelson NZ.

The Case 550E was built between 1991 and 1994. It was powered by a Case I 4-390, four cylinder, diesel engine which produced a SAE net hp of 67. It had four forward speeds of 2.5 to 5.8mph and four reverse from 2.7 to 6.4mph
Operating weight was 13,425lbs, length 12 ft 11 inches, width 96 inches and height 8 ft 4½inches.

From Case Corp. sales literature.

Crawler Tractor Scrapbook Pt 2   Case

This Case 750 was photographed in a yard at Gunnedah, NSW, Australia in 1993. It had a Case, four cylinder diesel engine rated at 1,900rpm with a bore and stroke of 4 x 4 1/8 inches. Dbhp was 36.25 maximum. Weight was 14,035lbs. Four forward speeds ranged from zero to 6.24mph

This Case 850 B belonged to Ernie Schicker, Whakatane district, NZ when photographed in 1999. It was just starting off on the afternoon's work clearing scrub and bush from steep hillsides. Although in the middle of summer it was soon bogged down in a gully.

The Case 850 B was made from 1976 to 1981. It had a Case A336BD diesel engine giving 72 engine hp. It had a powershift transmission and weighed 15,778lbs.

Crawler Tractor Scrapbook Pt 2 — Case

The **Case 1150** crawler was powered by a Case six cylinder diesel engine with a bore and stroke of 4 1/8 x 5 inches giving a maximum flywheel hp of gross 105, net 85 at 1850rpm. Weight was 15,357 lbs. Speeds forward were from zero to 6mph and reverse zero to 7.1mph. It was built in various forms from 1965 to 1996.

This early Case 1150 belonged to Malcolm Lumsden, Ohinewai, NZ when photographed in 1998.

Middle photo is a much later Case 1150 sitting in the Whakatane, NZ district in 1999.

Terratrac 500 had a Continental FA162, 4 cylinder, petrol engine or a GD157 4 cylinder diesel engine. Weight was 5310lbs (petrol) and 5470lbs (diesel), height 63.5", and length 101". Speeds were three forward up to 4.88mph and one reverse to 2.17mph. Made from 1957 to '63.

Terractrac 600 had a Continental F209, 6 cylinder petrol engine or a ED208 diesel engine. Shipping weight was 7,145lbs (petrol) or 7,205lbs (diesel). Length was 101", height 65". Speeds were 4 forward to 5.68mph and four reverse to 6.5mph. Made from 1957 to '62.

Terratrac 500   Terractrac 600

Crawler Tractor Scrapbook Pt 2 — Case

The **Case 1450** loader has crept in here because the author could not find a Case 1450 crawler except ones being wrecked for parts. As far as is know there were no Case 1550 crawlers in New Zealand.
The Case 1450 was powered by a Case A504BDT six cylinder, turbocharged, diesel engine giving a gross SAE horsepower of 144 at 2,100rpm or net 130. Height was 10.1 inches, weight was 33,680lbs. The transmission was Case's four speed powershift forward and reverse. The 1450 was built from 1979 to 1980 and the 1450B from 1980 to 1989. The Case 1450 had an increase in power to 140 netthp. The Case 1550 appeared in 1987.

Below the Case 1450 controls.   From Case Corp. sales literature.

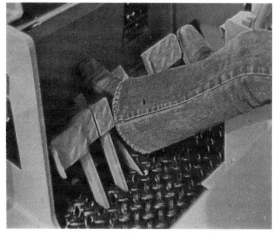

Crawler Tractor Scrapbook Pt 2

Case

When Case bought the American Tractor Corporation they inherited an innovative tradition. ATC were the first to have a powershift transmission in a crawler. In the mid 1950s a company was commissioned to design a powershift transmission with a torque converter and this appeared in their model 600.

The system was improved again with the 800 and 100 models having a choice of three ways to steer. There was the usual brake turn, the power turn and the counter-rotation turn.

After the Case takeover the transmission design was improved yet again, and the new design appeared in the 450, the 850 and the 1150 models.

The Case 1150 crawler steering instructions state the machine can be steered in four ways.

Gradual Turn: To turn left shift the left hand direction control lever to the neutral position, to turn right use the right lever.

Power Turn: Use the track speed control levers. To turn left shift the left hand track speed control lever to low and shift the right hand track speed control lever to high. To turn right shift the right hand lever to low and the left to high.

Privot Turn: A pivot turn is made by using the foot brakes. To turn left depress the left brake pedal. To turn right depress the right brake pedal.

Counter- Rotation Turn:
The counter-rotation turn is particularly useful in confined areas as the tractor will turn in its own length.

Place both track speed control levers in the low position. To turn left, place the left control lever in reverse and the right control lever in forward. To turn right place the right lever in reverse and the left lever in forward. When the desired turn has been accomplished, place both direction levers in the forward or reverse position and continue operation.

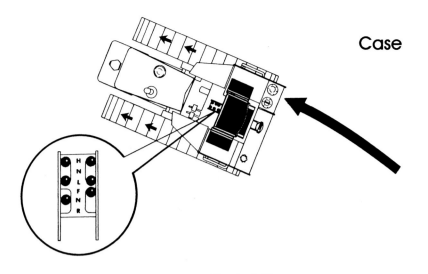

Gradual Turn

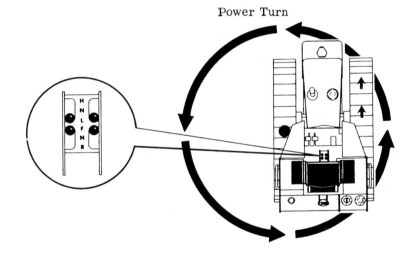

Power Turn

Pivot Turn

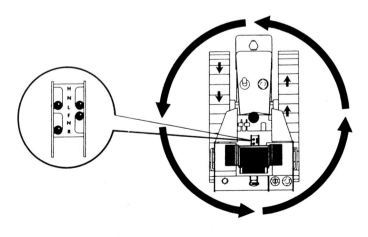

Counter-Rotation Turn

Crawler Tractor Scrapbook Pt 2

Case

Case 420

As reported earlier this year, John Fowler and Co. (Leeds), Ltd., a member of the Marshall organization, has arranged to make tractor components for J. I. Case Co., Ltd., who are tentatively entering the agricultural market with machinery (see elsewhere in this report): their tractor activities have so far been confined to the industrial sphere.

The latest Case tractor to be shown in Britain could, however, have agricultural applications. It is the medium horsepower 420 tracklayer. As exhibited, it was equipped with a hydraulically tilting dozer blade, 6 ft. 4 in. wide, and a rear toolbar with earth ripping tines. This machine, built in Racine, is intended solely for the industrial market so far: no drawbar models have been imported.

Originally petrol-engined, the 420 was withheld from the British market pending the development of a diesel version. The engine now fitted is a 188-cu.-in. unit developing 34 h.p. at the belt and 28 drawbar h.p. Specifications include controlled differential steering, 3-speed gearbox, and tracks of 4-ft. gauge having 14-in. plates and 4 ft. 9 in. length of ground contact.

Easy changeover...from tool bar to dozer operation

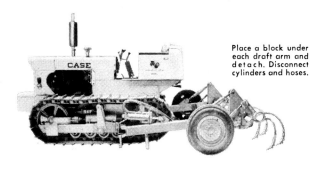

Place a block under each draft arm and detach. Disconnect cylinders and hoses.

Changing from tool bar to dozer, or vice versa, is a simple, one-man operation that takes little time and no lifting. You merely detach the draft arms and hydraulic connections at the sides of the tractor, turn the tractor around, then re-attach.

Drive forward enough to clear the draft arms, make a 180° turn, then head back in again.

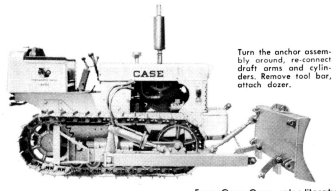

Turn the anchor assembly around, re-connect draft arms and cylinders. Remove tool bar, attach dozer.

From Case Corp. sales literature.

Crawler Tractor Scrapbook Pt 2 — Case

CASE Crawlers

Model	Production Start - Finish	Comments
CI	1929 -	Trackson, Roadless, Ajax conversions
LI	1929 -	Trackson, Roadless tracks
GT25, GT28, GT30, GT32,	1951 - 54	25hp, 4 cyl. ATC Terratrac 25dbhp, 30.4bhp
GT34, DT34,	1953 - 55	34hp, 4 cyl. diesel.
M2, M3, M3B, M4,	1952 - 59	
200, 256	1952 -	20ehp, 4 cyl. Continental F-140
300	1955 - 57	30ehp, 4 cyl. Continental F-140
310 diesel/petrol	1957 - 63	42ehp, 4 cyl. Case G188-D/ G148-B, 6,229lbs/ 5,939lbs
310C	1958 - 60	33ehp, 4 cyl. Case, 5,894lbs.
310D	1959 - 60	37ehp, 4 cyl. Case, 6,139lbs.
310E	1961 - 63	42ehp, 4 cyl. Case, 5,939lbs
310F	1963 - 64	37ehp, 4 cyl. Case, 5,900lbs
310G	1964 - 70	42ehp, 4 cyl. Case G148-B, 5,939lbs.
320	1957 - 59	? ehp, 4 cyl. Continental, 4,600lbs.
350	1963 - 79	39ehp, 4 cyl. Case G188-D, 7,971lbs.
350B	1979 - 87	39ehp 4 cyl. Case G188-D, 8,007lbs.
350C	1980 - 94	39ehp 4 cyl. Case 188-D, 10,099lbs.
356	1954 - 56	? ehp, Continental, 5,110lbs.
400	1955 - 56	40hp, 4 cyl. Continental F-162 petrol or diesel, 5,110lbs.
420	1957 - 70	36ehp, 4 cyl. Case G 148, 5,361lbs.
420B	1959 - 60	36ehp, 4 cyl. Case G148, 5,568lbs.
420C	1960 - 62	36ehp, 4 cyl. Case G148, 5,732lbs
450	1965 - 79	51ehp, 4 cyl. Case G188-D, 10,795lbs.
450B diesel/petrol	1978 - 84	53ehp, 4 cyl. Case G270D/G188, 9,229/8,440lbs
450C	1983 - 95	63ehp, 4 cyl. Case 4T-390, 11,549lbs
455B	1983 - 92	? ehp, Case G-207D 13,000lbs
455C	1984 - 91	? ehp, Case 4-390 14,249lbs
456	1983 - 92	
475	1972 - 87	81ehp, 4 cyl. Case 301BD, 14,616lbs.
475C	1978 - 87	81ehp, 4 cyl. Case 301BD, 15,430lbs
500	1957 - 63	? ehp, 4 cyl.. Continental FA-162 petrol, ED-208 diesel, 5,310lbs.
500B	1958 - 62	? 4 cyl. Continental, 5,310lbs
520 diesel/petrol	1959 - 60	43ehp, 4 cyl. Continental GD-157/ 162, 5,637/5,008lbs
550	1989 - 93	67ehp, 4 cyl. Case 4-390, 13,400
550E	1991 - 94	67ehp, 4 cyl. Case 4-390, 12,900lbs
556	1955 - 57	
600	1957 - 62	39ehp, 6 cyl. Continental F-209 petrol, ED-208 diesel, 6,901lbs. ??hp
600B	1958 - 62	39ehp, 6 cyl. Continental ED208, 7,201lbs. ??hp
610	1957 - 62	56ehp, 4 cyl. Cont. 208, Case-F-209 (6cyl.) Case ED-208 (4cyl.) 8,901lbs.
610B	1961 - 65	39ehp, 4 cyl. Continental 208CID, 9,201lbs.
610D Terra trac	1961 - 65	56ehp, 4 cyl. Continental ED208.
650	1989 - 91	80ehp, 4 cyl. Case 4T-390, 15,700lbs
650E	1992 - 96	80ehp, 4 cyl. Case 4T-390, 15,480lbs.
656 diesel/petrol	1955 - 57	? Case
730	1960 - 63	
740	1960 - 63	
750	1961 - 67	81ehp, 4 cyl. Case A267-D, 11,720lbs.
800	1957 - 61	40ehp, 4 cyl. Conintental HD-227, 11,210lbs.
800BR	1961	
800C	1959 - 61	40ehp, 4 cyl. Continental HD-227, 11,510lbs.
810	1957 - 61	40ehp, 4 cyl. Continental HD-277, 13,960lbs.
810C	1959 - 61	40ehp, 4 cyl. Continental HD-277, 14,260lbs.

Crawler Tractor Scrapbook Pt 2 — Case

Model	Years	Specs
830	1960 - 63	
840	1960 - 63	
850	1967 - 77	72ehp, 4 cyl. Case A301-BD, 17,382lbs.
850B	1976 - 81	72ehp, 4 cyl. Case A336-BD, 15,778lbs.
850C	1981 - 94	78ehp, 4 cyl. Case A336-BD, 17,027lbs
850D	1894 - 96	82ehp, 6 cyl. Case 6-590, 19,177lbs
850E	1992 - 94	89ehp, 6 cyl. Case 6-590, 16,800lbs.
855C	1984 -	78ehp, 6 cyl. Case A-336BD
855D	1984 - 91	
855E	19192 - 94	
930	1960 - 63	
1000	1957 - 65	90ehp, 6cyl.
1000BR	1961	90ehp, 6cyl.
1000D	1960 - 65	92ehp, 6cyl.
1010	1957 - 65	90ehp, 4cyl. Continental
1010C	1959 - 60	92ehp
1150	1965 - 75	105ehp, 6cyl. CaseA401
1150B	1973 - 80	105ehp, 6cyl.
1150C	1978 - 84	105ehp, 6cyl. Case A451BD
1150D	1983 - 94	110ehp, 6cyl. Case 6T-830
1150E	1985 - 96	113ehp, 6cyl. Case 6T-590, Long track Case 6-830
1155D	1983 - 86	
1155E	1985 - 92	
1200TK	1966 - 67	
1450	1973 - 80	130ehp
1450B	1980 - 89	140ehp, 6cyl. Case 504
1455B	1983 - 87	140ehp
1550	1987 - 96	150ehp, 6cyl. Case 6T-830 hydrostatic transmission

E & OE

Case Terratrac 800
Tilt-Crown Dozer had a maximum drawbar hp of 40.41form a 4 cylinder diesel engine with a bore and stroke of 4" x 5.5". It had four forward speeds up to 6mph and four reverse up to 7mph. Height was 68", width 69", length 113½" and weight 13,430lbs with blade.

Case Terratrac 1000
Angling Dozer was made from 1957 to '65. It four cylinder diesel engine had a bore and stroke of 4½" x 6" giving a maximum drawbarhp of 55.51. Speeds were the same as Terratrac 800. Height was 68", width 76", length 113½" and weight 15,700lbs with blade.

FIAT

Fiat was founded by Giovanni Agnelli in 1899. It was created to make motor-driven vehicles and was named Fabrica Italiana di Automobili Torino, today it is known as Fiat.

In 1919 Fiat began the production of its first wheel tractor, the 702. In 1932 the first Fiat crawler tractor appeared, the 700C, which remained on the market until 1940. This was Europe's first mass produced crawler tractor.

Quality fuel being difficult to obtain during the 1930s the Fiat designers came up with a multi fuel engine. This came out in the Model 40 Boghetto crawler tractor in 1938. The engine produced 41.5hp and could run on diesel fuel, alcohol, low-octane gases, natural gas, petrol or fuel oils.

Fiat's first diesel engine crawler tractor came out in 1946, the Model 50.

By 1972 Fiat crawler tractors were being made at Modena and Lecce in Italy, and assembled in Australia, Brazil, Japan, Morocco, and Zaire. There were 14 basic crawler tractors plus special versions.
In 1974 Fiat bought into Allis Chalmers to form **Fiatallis**, All Fiat machinery that worked on the land, ie agricultural and construction, was brought together in 1988 under the name of Fiat Geotech. Fiat is now involved with Fiat-Hitachi.

The Fiat 40 was built from 1939 to 1947. There was a Mark 3 and a Mark 4. They were powered by a 4 cylinder petrol engine with a bore and stroke of 95 x 140mm producing 41.5hp at 1,500rpm. The Mark 3 had three forward speeds from 2.4 to 11km/h and reverse and the Mark 4 had four forward speeds from 2.4 to 8.5km/h.

The first Fiat crawler and the first European crawler to be mass produced was the Model 700C. It had a four cylinder petrol engine with a bore and stroke of 90 x 140mm, producing 30hp. Speeds were three forward and one reverse.. It was built from 1932 to 1940.

Fiat 40 1939

Fiat 40 1946

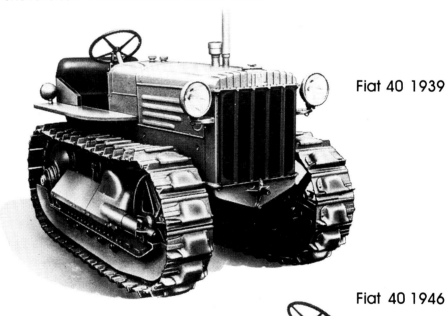

Illustrations from Giovanni Magnanini's collection.

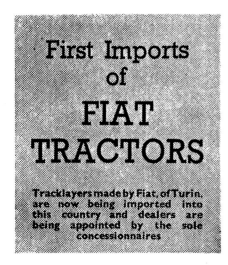

First Imports of FIAT TRACTORS

Tracklayers made by Fiat, of Turin, are now being imported into this country and dealers are being appointed by the sole concessionnaires

November-December, 1950   FARM MECHANIZATION

Model 52 ploughing. The steering wheel is an unusual feature on a tracklayer.

FIAT tractors are now on the British market. Made by the well-known Italian firm, Fiat, of Turin, the first shipment arrived in England during August. Sole concessionnaires for the United Kingdom are MacKay Industrial Equipment, Ltd., Faggs Road, Feltham, Middlesex, who have many years' experience in distributing and servicing industrial tractors.

Fiat tractors have proved themselves on the Continent to be sound machines. Twelve have been on trial in this country during the past eighteen months and good reports have been received by MacKay's of their performance on Agricultural Committee contract work. It is planned to import Models 50, 52 and 601.

Different Track Gauges

Models 50 and 52 differ only in the gauge of the tracks and in the gearbox ratios. Model 50 is the narrower of the two, with a gauge (centre to centre) of 3 ft. 9 ins., overall width of 5 ft. 3 ins., a track-shoe width of 1 ft. 1¼ ins., a bearing surface on soft ground of 1,862 sq. ins., a weight in running order of approximately 9,900 lb., and pressure on soft ground of 5½ lb. per sq. in.

Model 52 has a gauge of 4 ft. 11 ins., overall width of 6 ft. 4¾ ins., a standard track-shoe width of 1 ft. 5¾ ins., a bearing surface on soft ground of 2,395 sq. ins., a weight in running order of approximately 10,560 lb., and ground pressure of approximately 4½ lb. per sq. in. (For work on very soft ground, ground pressure can be still further reduced to a minimum of about 2⅜ lb. per sq. in. by fitting special track shoes, 2 ft. 1½ ins. wide.)

Both models are powered by a Saurer four-cylinder, four-stroke, direct-injection Diesel engine, with bore and stroke of 110 × 140 mm., a total piston displacement of 5,320 c.c., and a maximum b.h.p. of 50 at 1,400 r.p.m. (governed at full load). Belt h.p. is given by the manufacturers as 45 and drawbar h.p. as 38 (a slightly higher db.h.p figure can be expected of Model 52 owing to its wider tracks and lower gearbox ratios).

The Diesel engine is started by means of a 10 h.p. horizontally opposed twin-cylinder petrol engine. The single-plate dry clutch can be removed and replaced without disturbing the engine. Steering is by clutch and brake, the steering clutches being of the multiple-disc type actuated by a steering wheel instead of the more usual clutch levers. There are two foot brakes which act on the steering clutch drums, and one hand brake.

Oscillating Track Frames

The track frames are fully oscillating and shocks are absorbed by two heavy-duty transverse leaf springs anchored at their centres and supported by hardened-steel rollers at their outer ends. All support rollers, track wheels and front idlers are fitted with roller bearings, protected by dirt-excluding seals. The track itself is of the orthodox twin-rail, pin-and-bush type which, for road travel, can be fitted with plates which are spring-fastened and clipped into position with much greater ease than screw-on plates. A prising bar is supplied for plate removal.

Features of the engine are the toroidal form of the combustion chambers and the fact that there are two inlet and two exhaust valves per cylinder. Direct injection is by a Bosch type of pump and Saurer fuel injectors, the design of which, it is claimed, makes incorrect assembly impossible.

Another refinement is a metal-edge type of lubricating-oil filter, the centre axle of which is connected by linkage to the main engine clutch, so that every time the latter is operated, the axle is turned so that dirt between the laminations falls to the bottom of the filter, from which it can be removed when the assembly is dismantled for cleaning. A dual-geared lubricating-oil pump is provided, so that full lubrication takes place when the machine is working on steep gradients. Cooling is by centrifugal pump, four-bladed fan and monobloc radiator.

Spare Injectors

Model 50 is priced at £2,200 and Model 52 at £2,300. Both prices are ex works (England) and include a 6-volt electric lighting equipment with 75-watt dynamo, hour meter, grease gun, four spare injectors, and a set of tools.

Extras include road plates, rear-mounted belt pulley (diam. 11¾ ins., face 7⅛ ins., speed 815 r.p.m.), power take-off (shaft speed, 550 r.p.m.) and fuel filtering pump for tank filling.

It is expected that the greatest demand in the United Kingdom will be for the wider-gauge tractor (Model 52). Model 601 is a small tracklayer with dimensions suitable for orchard work, powered by a petrol/paraffin engine.

Service Arrangements

Dealers are being appointed by MacKay's, who will maintain a supply of parts to assist dealers in after-sales service. A "Farm Mechanization" representative was informed by the concessionnaires that, so far as possible, their organization for servicing industrial tractors will be applied to aid agricultural machinery dealers in Fiat service.

This photograph shows the different track gauges of the Model 50 (left) and the Model 52.

Crawler Tractor Scrapbook Pt 2 — Fiat

The Fiat 25C had a maximum belt hp of 23 and drawbar hp of 18 produced by a kerosene fueled, 4 cylinder engine. Bore and stroke were 85mm & 100mm, rpm was 1,750. Its weight was 3,3380lbs. Speeds were 2.1, 2.9, 4, 6.7mph forward and 2.2 mph reverse. It was 44½ inches high, 48 inches wide and 92 inches long.

This machine belonged to Jim Richardson, Edgecombe, NZ when photographed in 1998.

The Fiat 25CI came out in 1954 with a diesel engine. Its details appear on page 44.

This 1952 Fiat CI belonged to David Hunger when photographed in 1994 at New Plymouth, NZ.

Serial number was 600/4045.

The Fiat 40C or OM 35/40 was built between 1952 and 1960. It was powered by an OM COD/40 direct injection diesel engine with a 100 x 120mm bore and stroke.

Photographed in the Nelson district, NZ, by Bill Higgs in 1999.

Crawler Tractor Scrapbook Pt 2 — Fiat

Production of the Fiat 311C or 351 Hillside model was started in 1959. It was powered by a 27 belt hp, four cylinder engine with bore and stroke of 3.228 x 3.543 inches. Forward speeds were .9, 2.2, 3, 4, and 6.7 mph and reverse 2.1mph. Total weight was 4,020lbs. Overall width of the hillside model 58 inches.

This Fiat 351 was photographed by Jim Richardson in 1987.

Below: Ernie West stands beside his Fiat 70C in Taranaki, NZ in 1999. The Fiat 70C came on the market in 1961. It was powered by a Fiat 604-035 diesel engine which gave 74ehp and 58.3 dbhp.

June, 1951 — FARM MECHANIZATION

THE FIAT TRACTOR
Model 55

Manufactured in Italy, the Model 55 is the latest addition to a range of Fiat tracklayers distributed throughout Great Britain by Mackay Industrial Equipment, Ltd., Faggs Road, Feltham, Middlesex

CONFORMING to well-known principles of tracklayer design, the Fiat specification includes Diesel engine; clutch and brake steering; "A" frame suspension and steel tracks. The Model 55 weighs 11,130 lb. and has a maximum drawbar pull of 12,125 lb.

The 4-cylinder engine develops 55 brake horse-power at the maximum governed speed of 1,400 r.p.m. Starting is by a 10-brake-horse-power petrol engine, which operates on the main engine flywheel. Other main features are hand-operated over-centre clutch and 5-forward and one reverse-speed gearbox. Provision is made for belt pulley, winch, and heavy earth-moving equipment.

Cooling System

The cooling system is pump-assisted thermo-siphon and includes belt-driven four-blade fan, impeller, thermostat and a radiator of bolted construction. A dynamo is included in the fan belt circuit and its mounting provides for belt tension adjustment which is correct when there is approximately ½-in. slack between the dynamo and the fan pulleys.

Renewable wet cylinder liners are used. Each liner is supported on the crankcase by a flange integral with the liner. Shims are provided between each flange and its seat to enable the liners to be adjusted to the cylinder-head level. The maximum permissible wear of the liner bore diameter is given as 0.006 in.

Aluminium-alloy pistons are provided with two compression and four scraper rings. The correct gap for all rings is between 0.012 and 0.020 in. Pistons and connecting rods can be removed from the top of the liners. The crankshaft is carried on five main bearings. These, and the big-end bearings, are the renewable precision shell type. Crankcase inspection doors make it possible to renew the big-end bearings without removing the sump.

The engine lubrication system includes a twin gear pump. Two pump feed pipes are provided. One picks up from the rear of the sump, the other from the front, so that the pump remains supplied irrespective of the angle at which the engine may be called upon to work.

From the pump the oil is supplied at a pressure of 42 lb. per sq. in. to the crankshaft, camshaft, timing gears and overhead valve assembly. Two filters are included in the pressure circuit. The first embodies a self-cleaning element which is coupled to the master clutch control lever. Dirt from this filter falls into a sludge trap which should be drained every 60 working hours. The second filter is a "Fram," mounted on the right-hand side of the cylinder block. Its element should be renewed every 1,200 hours.

An engine oil cooling radiator—useful in the tropics—is available as optional equipment. It fits onto the front of the radiator and is pressure fed from the main supply pump. One of the most valuable accessories supplied as standard equipment is a mechanically operated hour meter mounted on the left-hand side of the engine, and gear driven from the camshaft.

The electrical system needs no batteries as it provides direct lighting from a 6-volt 90-watt dynamo. The dynamo drive incorporates a hand-operated clutch which enables the dynamo to remain idle when not required. Two headlamps and one rear floodlight are standard equipment.

Air Cleaner and Fuel System

An oil-bath air cleaner is provided. Every 20 working hours the oil should be inspected and must be renewed if containing more than ⅜-in. depth of sediment. Oil level is critical in the function of the cleaner. The correct level is shown by a mark on the bath. If the level is found to have dropped below this mark after the engine has been shut down for not less than 30 minutes, the cause must be rectified: either the oil will be too thin or the inlet pipe partially choked. The correct grade of oil is that recommended for the engine (see specification at end of article). In dusty conditions the air cleaner elements should be removed and washed in clean fuel every 120 working hours.

The injection pump is a Bosch type and incorporates centrifugal governor, excess-fuel starter aid, and suction lift pump.

(Continued overleaf)

Left: In addition to a "Fram" oil filter (A), a self-cleaning filter is embodied in the timing gear cover. Its cleaning mechanism is actuated by the main clutch lever through connection B.

Crawler Tractor Scrapbook Pt 2

Fiat

The Fiat 55 (contd.)

From the tank, fuel is forced through a renewable element filter by the suction-lift pump which is mounted on the side of the injection pump. The element should be renewed every 1,200 working hours. A hand primer is incorporated in the suction-lift pump to enable the system to be primed and bled after element renewal. Lubrication of the injection pump and governor is by splash from an oil sump provided by the pump base.

The Starting Engine

The Diesel engine is started on full compression by a petrol engine mounted on the flywheel housing. This engine is a horizontally opposed, two-cylinder, side-valve unit with a piston displacement of 40.5 cubic ins. It develops 10 brake-horse-power at 3,000 r.p.m. and is fitted with high-tension magneto, gravity-fed Weber carburetter, fabric air cleaner and a crankcase breather (to be closed when the engine is not in use).

Lubrication is by splash from a wet sump independent of the Diesel engine, but the cooling system takes its water from the main cooling system, so that heat from the starter engine is transferred to the Diesel engine. This facilitates starting in cold weather.

Maximum starter engine speed is controlled by a centrifugally operated primary circuit cut-out embodied in the magneto. A cord is wound round a grooved flywheel and pulled to start the engine. The flywheel must be turned clockwise as viewed from the driver's seat.

The starter engine engages a ring gear on the Diesel engine flywheel. The mechanism includes three constant-mesh gears, a clutch, brake, splined shaft, sliding pinion and two control levers. The first of the constant-mesh gears is keyed to the starter engine crankshaft; the second transmits the drive from this gear to the third gear which is free to turn on the splined shaft that also carries the clutch, brake and sliding pinion.

The sequence of operation is as follows: when the starter engine is running, the constant-mesh gears turn. Movement of the vertical control lever towards the rear of the tractor disengages the clutch and applies the brake and thus holds the splined shaft stationary against normal oil drag of the third constant-mesh gear. The horizontal lever is then moved upwards to engage the sliding pinion with the flywheel ring gear. Two spring-loaded latches lock the pinion in engagement. The vertical lever must then be moved forward to engage the clutch which transmits the drive from the third gear to the starter pinion. When the Diesel engine starts, centrifugal action automatically disengages the sliding pinion.

The main point to observe is that the brake must be applied before the sliding pinion control lever is moved to its engaged position, otherwise the pinion may turn against the ring gear to the detriment of both when being engaged. Should the sliding pinion fail to disengage when the Diesel engine has started, the clutch must be released to disconnect the drive.

The clutch is adjustable by means of a spring-loaded latch which is accessible when the control-lever support base has been removed. Correct adjustment provides for an effort of approximately 30 lb. to engage the control lever. The sliding pinion-disengaging mechanism is adjustable by screws situated in the rear ends of the engagement latches.

Engine Clutch

From the flywheel, the drive is transmitted to the gearbox by a hand-controlled over-centre clutch. The driving plate is bolted to the flywheel rim and operates between a pair of plates, one of which is fixed to the clutch shaft, and has a serrated hub along which the opposing plate may slide. When the control lever is pulled rearwards, a cam ring pivots four arms which then exert pressure on the sliding plate and thus engage the clutch. A brake is incorporated in the throw-out mechanism to operate on the clutch shaft when the clutch is fully disengaged.

One lubrication point and one adjuster are provided. The lubrication point supplies the clutch throw-out mechanism. It is situated on the top of the clutch housing and should be filled every 20

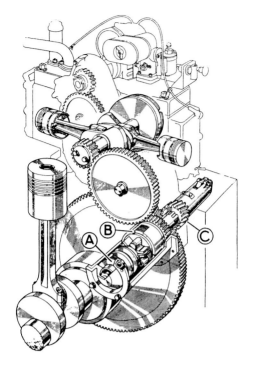

Starter engine engaging mechanism. (A) clutch shaft brake, (B) clutch and (C) sliding starter pinion shown engaged to main flywheel.

hours. The adjuster is situated towards the centre of the clutch and varies the pressure that can be exerted on the sliding plate by the engagement arms. When it is correctly set, the lever will engage with a distinct snap under a force of approximately 50-55 lb. The clutch can be removed without disturbing the engine and gearbox.

To adjust the clutch remove the clutch inspection cover, disengage the clutch and turn the pressure plate spider until a spring-loaded plunger in the spider hub becomes accessible. Engage any gear, withdraw the plunger and turn the spider clockwise to tighten the clutch or anti-clockwise to slacken it.

The gearbox is formed by the main transmission housing casting and conforms to orthodox design. It has an oil capacity of approximately 4½ gallons. This oil also lubricates the bevel gear drive. The correct level is denoted by a plug on the right-hand side of the housing.

A bevel pinion and crown gear transmit the drive from the gearbox to two multi-plate steering clutches which are controlled by a steering wheel. A feature of the steering is that when the wheel is moved to its extreme right or left position it automatically holds the corresponding clutch disengaged, so that the driver's hands are both free for operating implement controls, etc. A contracting band brake operates on each clutch drum and is controlled by a foot pedal. Both brakes are connected to a single lever for

Left: (A) dynamo-drive clutch lever, (B) hour meter, (C) starter engine clutch and pinion control levers, (D) left-hand cylinder head of starter engine.

June, 1951   FARM MECHANIZATION

Left: Components of track assembly. The pins and bushes should be turned when track adjustment has allowed a 3-in. forward movement of the adjuster yoke.

The instrument panel carries oil-pressure and water-temperature gauges, switch, dashlight and fuse-box. The steering clutches are wheel controlled. Independent pedals actuate the brakes, which are also connected to a lever for parking use.

parking. Independent adjusters are provided for each clutch and brake. The final drive to each track sprocket is through spur reduction gears.

Suspension

"A" frame suspension allows independent vertical oscillation of the track frames. The forward weight of the tractor is carried on the frames by a laminated transverse spring interconnected with a rebound spring. The rear weight is carried by dead axles around which the frames are free to pivot. Lateral stability is provided by the "A" frame which extends diagonally from each track frame to the centre of the dead axles.

The bottom rollers and front idlers are carried on roller bearings. Each front idler is spring-loaded to absorb track shock recoil. The track adjustment mechanism is also incorporated in the front idler yoke. The correct track tension allows the track to be lifted 1½ ins. above the top carrier roller.

Induction-hardened pins and bushes are press-fitted into steel-track links. The life of the track chain and sprocket will be prolonged considerably if the pins and bushes are turned when adjustment has allowed the front idler yoke to move forward 3 ins. The measurement should be taken from the rear of the yoke guide (see 32 on the sectional drawing of the Fiat 55 overleaf).

Service

After-sales service facilities offered by Mackay's through their dealer organization include the supply of reconditioned sub-assemblies ranging from a track roller to a complete engine. A service school has also been instituted for dealer service personnel.

Fiat Dealers

Dealers appointed by Mackay's are: Fredk. H. Burgess, Ltd., The Green, Stafford (subsidiary companies and all branches); H. A. Collings, Tractor Depot, Biggleswade, Beds; Collisons (Beverley), Ltd., Norwood, Beverley, Yorks; Cowlishaw and Sons, Methwold, Thetford, Norfolk; E. Doe and Sons, Ltd., Ulting, nr. Maldon, Essex; Eastern Roadways (Engineers), Ltd., Stansted Road, Bishops Stortford, Herts; D. T. Gratton and Sons, Ltd., Wide Bargate, Boston, Lincs.; K.E.F. (Engineers), Ltd., Phœnix Works, Tovil, Maidstone, Kent; J. Mann and Son, Ltd., Saxham, Bury St. Edmunds, Suffolk; Power Farming, Ltd., Casterton Road, Stamford, Lincs; Neil Ross, Nor'East Tractor Works, Bridge Street, Ellon, Aberdeenshire; John Rutherford and Sons, Ltd., Home Place, Coldstream, Berwicks; Shorwell Plant, Ltd., Bowcombe, Carisbrooke, Newport, Isle of Wight; Alex Strang (Tractors), Ltd., Pipe Street, Portobello, Midlothian; Sun Engineering (Crowle), Ltd., Crowle, Scunthorpe, Lincs.; Tractors and Motors (Southern), Ltd., Avenue Road, Brockenhurst, Hants; Watkins and Roseveare, Ltd., Cantrell Works, Ivybridge, S. Devon.

ABRIDGED SPECIFICATION OF THE FIAT 55

General description. — Clutch- and brake-steered Diesel-engined tracklayer. Provisional price, £2,600.

Engine.—Fiat four-cylinder vertical four-stroke. Bore, 4.703 ins.; stroke, 5.512 ins.; swept volume, 399.4 cubic ins. Compression ratio, 15 to 1. Brake horse-power, 55 at maximum governed speed of 1,400 r.p.m. Maximum torque, 224 lb./ft. at 900 r.p.m.

Crankcase and cylinders. — Monobloc. Renewable wet liners. Fire-bearing crankshaft. Steel-backed lead-bronze shell bearings. Tappet clearances (cold), inlet 0.012 in., exhaust 0.016 in.

Lubrication system.—Pressure feed from gear-type pump. Oil filters, two—"Autoclean" and "Fram."

Air, fuel and injection equipment.—Oil-bath air cleaner. Bosch-type multi-element injection pump with mechanical governor. Four-hole injector nozzles. Injection, direct at 3,237 lb. per sq. in.

Starting.—By horizontally opposed, two-cylinder, four-stroke, 10 b.h.p. petrol engine. Bore, 2.953 ins.; stroke, 2.953 ins.; total swept volume, 40.5 cubic ins. Tappet clearances (cold), inlet 0.010 in., exhaust 0.012 in. Contact-breaker gap, 0.012-0.016 in.

Electrical equipment.—Direct lighting from 6-volt, 90-watt dynamo. Two head lamps, one rear floodlight.

Main engine clutch.—Dry single plate, hand operated, over centre.

Gearbox.—Five forward speeds, one reverse.

	M.p.h.	Maximum drawbar lb. pull at maximum torque
1st	1.4	12,125
2nd	2.3	7,825
3rd	2.8	5,840
4th	3.6	4,740
5th	5.2	—
Reverse	1.8	—

Steering.—Multiple-plate clutches and contracting band brakes.

Tracks.—Steel links and press-fitted pins and bushes. Clip-on road pads.

Suspension. — "A" frame. (Allows independent vertical movement of track frames.) Two dead axles extending through sprocket hubs support rear weight on rear of track frames. Forward weight carried on frames by transverse laminated spring interconnected with laminated rebound spring. Alignment maintained by "A" frame bracing.

Other Details

Weight	11,130 lb.
Track centre	59 ins.
Track plate width (standard)	17¾ ins.
Total area of ground contact	2,400 sq. ins.
Ground pressure	4.7 lb. per sq. in.
Overall length	121 ins.
Height to bonnet top	61½ ins.
Minimum ground clearance	11¼ ins.

Capacities.—Radiator, 10½ gallons; sump, 4 gallons; gearbox and bevel gear housing, 4½ gallons; final-drive reduction housings (each), 2½ gallons.

Lubrication Recommendation.—Diesel engine: S.A.E. 50 above 95 degrees F., S.A.E. 40 above 50 degrees F., S.A.E. 30 below 50 degrees F., and S.A.E. 20 below 5 degrees F. (renew every 60 hours). Starter engine: S.A.E. 10 (renew every 1,200 hours). Gearbox and bevel gears: S.A.E. 140 (renew every 1,200 hours). Reduction gear housing: S.A.E. 90 (renew every 1,200 hours).

Sole concessionaires for Great Britain; Mackay Industrial Equipment Ltd., Faggs Road, Feltham, Middlesex. Phone: Feltham 3435-9.

The Fiat 55

A drawing specially prepared by FARM MECHANIZATION

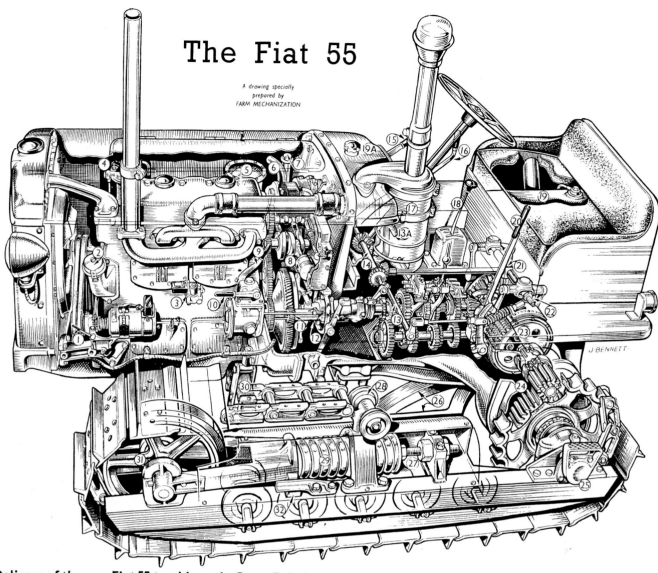

Delivery of the new Fiat 55 tracklayer in Great Britain is starting this month. The tractor weighs 11,130 lb. and has a maximum drawbar pull of 12,125 lb. at maximum torque. The engine develops 55 brake h.p. at maximum governed speed of 1,400 r.p.m.

KEY TO NUMBERS ON DRAWING

1. Dynamo drive clutch lever.
2. Oil filler cap, main engine.
3. Hour meter.
4. Full-flow fuel filter.
5. "Fram" lubricating-oil filter.
6. Starter-engine exhaust pipe.
7. Starter-engine breather.
8. Starter-engine crankshaft assembly.
9. Starter-engine clutch lever.
10. Starter-engine pinion engagement lever.
11. Starter-engine pinion (engaged).
12. Main clutch throw-out assembly.
13. L.H. steering-brake cross-shaft.
13a. L.H. steering-brake pedal.
14. Steering-clutch control cross-shaft.
15. Throttle control lever.
16. Main clutch lever.
17. Oil-bath air cleaner.
18. Gear-change lever.
19. Main fuel tank.
19a. Starter-engine fuel tank.
20. Parking brake lever.
21. Steering-clutch operating cams.
22. Steering-clutch adjuster screw (brake adjuster nut immediately below).
23. L.H. steering-clutch assembly.
24. L.H. final-drive reduction gear.
25. Track frame axle cap.
26. L.H. track "A" frame.
27. L.H. track adjuster screw.
28. Track support roller.
29. Track recoil springs.
30. Transverse springs, main and damper.
31. Adjustable front idler.
32. Front-idler yoke guide.

AUG. 1, 1952 FARM IMPLEMENT AND MACHINERY REVIEW

The New "Fiat" Model "35" Crawler Tractor

PRELIMINARY details and a preview of the new "Fiat" model "35" crawler tractor were obtainable at the stand of **Mackay Industrial Equipment, Ltd.**, Faggs Road, Feltham, Middlesex, and it is hoped to be able to give deliveries early in the spring of 1953. The machine is powered by an "O.M." overhead valve 4-cylinder, 4-stroke diesel engine that gives a belt h.p. of 35, and a drawbar h.p. of 29, while there is a maximum drawbar pull of 7,716lb. Direct fuel injection is employed, and there is a "Bosch" type pump used in conjunction with 4-hole injectors. The piston crown incorporates double turbulence cavities as an aid to efficient combustion, and wet liners are used in the cylinders, so that maintenance and replacement costs are low. A centrifugal-type governor is normally set at 1,550 revs. per min., and engine starting is by a 24-volt electric starter. A twin-gear oil pump driven off the crankshaft is used for engine lubrication and this system incorporates two oil-filters, the primary being a self-cleaning, edge-type, which is rotated ⅛th of a turn each time the master clutch is operated, and the secondary is of the replaceable cartridge design. Of the two air-cleaners fitted, a pre-cleaner removes 60 to 70 per cent. of the heavy matter, and an oil-bath cleaner gives the final purification. A water impeller is used in conjunction with a 4-bladed fan for engine cooling and adjustable radiator shutters are fitted. Steering is by hand levers operating dry multi-plate steering clutches, the drums of which are used as braking surfaces for the hand and foot-operated brakes. The master clutch is a dry single-plate unit of the over-centre type and can be withdrawn without disturbing the engine. A transverse main spring is used for suspension and stability. Four double-flanged truck rollers are used on each track, are mounted on taper roller bearings, and require lubrication only every 300 hours. The electrical system incorporates two 12-volt, 60-amp. hour batteries coupled in series to give a 24-volt supply, and a dynamo with cut-out controls is used for battery charging.

* * * *

"Fiat" Crawler Tractor Model "O.M." 35/40C

Made by O. M. Societa per Azioni, Milan, Italy. Date of Test:—July, 1953. Report No. BS/NIAE/53/11.

APRIL 1, 1954

BRIEF SPECIFICATION

Fuel:—Diesel oil; specific gravity 0.830 at 60deg. fah.; Cetane No. 56. Nearest U.S. equivalent: commercial diesel fuel.

Tractor:—Serial No. 50115.

Engine:—"O.M." model COD/40 direct-injection diesel engine, serial No. 001667; four cylinders, vertical, in-line; 100mm. bore by 120mm. stroke, compression ratio 16:1; overhead valves; renewable wet cylinder liners; "O.M."-type KC60R1F injectors fed by "O.M." type PE4B80E410L4/8 injection pump; "O.M." plunger-type fuel feed pump: sediment bowl and cartridge-type fuel filter; "O.M." centrifugal governor, governed range of engine speed 500 to 1,500 revs. per min.; forced-feed lubrication from gear-type pump, self-cleaning filter; "O.M." oil-bath air-cleaner with pre-cleaner; centrifugal pump cooling system with 4-blade, 16½in. dia. fan; thermostat and radiator shutters for temperature control; two 6v. "Marelli" lead-acid batteries: 12v., 60 amp./hour system for starting and lighting; electrical starter motor and excess fuel device for cold starting; fuel capacity, 11 gall., oil capacity, 17½pt., cooling water capacity, 4gall.

Transmission:—11.2in. dia. "O.M." single-plate, dry clutch: hand-lever operated; six forward speeds and two reverse; oil capacities: gearboxes (main and transfer), 37pt., final drives, 7½pt. Steering by hand-lever operated multi-plate steering clutches and foot-pedal operated external-contracting band brakes: brake locking lever and ratchet for parking. Pin-jointed tracks: 1.2in. dia. pins, 32 per track; track pitch, 6.3in.; 12.2in. wide trackplates with integral grousers 1.7in. deep; driving sprocket of pitch dia. 27.3in. and 2in. face width with 27 teeth; four bottom track rollers per track at 9in. centres, rolling dia. 7.1in.; one top carrier roller per track, rolling dia. 5.5in.; front idler wheel of 21.6in. rolling dia.; distance between driving sprocket and front idler wheel centres: minimum 4ft. 9½in., maximum 4ft. 11½in.; track gauge 43.3in.; approximate length of track in ground contact: 4ft. Suspension by track frame pivoted about dead axle at rear and mounted on transverse leaf spring at front. 6-spline, 1⅜in. dia. power take-off shaft; 9.45in. dia. by 6.9in. face width belt pulley.

Drawbar:—Swinging drawbar, radius of swing 33.6in., pivot centre 21.7in. forward of sprocket centre. Position of drawbar and power take-off shaft as specified in B.S. 1495/1948, A = 9½in., B = 4¾in. to rear, C = 11½in., D = 10½in., c = central.

Nominal speeds (at 1,500 revs. per min. rated engine speed):—Low ratio: L1 gear 1.04, L2 1.66, L3 2.36 miles per hour. High ratio: H1 gear 2.84, H2 4.57, H3 6.49 miles per hour. Reverse: low 1.36, high 3.73 miles per hour. Belt pulley, 1,260 revs. per min., power take-off, 600 revs. per min.

Weight:—Total weight of tractor (including fuel, oil, cooling water, belt pulley and operator: weighbridge figure) was 7,540lb.

Dimensions:—Overall length of tractor, 9ft. 1in.; overall width, 4ft. 10in.; overall height, 6ft. 4in. (to top of exhaust).

FARM IMPLEMENT AND MACHINERY REVIEW OCT. 1, 1954

Diesel Engined "Fiat"

"25 CI" In More Economical Form But at the Same Price

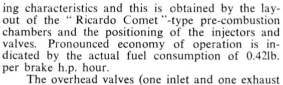

A Side View of the New "Fiat" Crawler

It is announced by Mackay Industrial Equipment, Ltd., Faggs Road, Feltham, Middlesex, that the "Fiat" model "25 CI" crawler tractor is now available on the British market in a more powerful form, in that, in place of the former petrol/paraffin engine, it is fitted with a "Fiat" diesel type without any alteration in price.

This diesel engine has a bore of 3.22in. by a stroke of 3.54in. and a capacity of 1901c.c., with the compression ratio 20:1. A 4-cylinder, 4-stroke type, developing 25 h.p. at 2,000 revs. per min. and giving a drawbar h.p. of 20 and a belt h.p. of 24, the engine has smooth running as one of its outstanding characteristics and this is obtained by the layout of the "Ricardo Comet"-type pre-combustion chambers and the positioning of the injectors and valves. Pronounced economy of operation is indicated by the actual fuel consumption of 0.42lb. per brake h.p. hour.

The overhead valves (one inlet and one exhaust per cylinder) are operated through tappets, pushrods and rockers by a 3-bearing camshaft driven by helical gears from the front of the engine. A "Fiat"-made "Bosch"-type 4-element block pump delivers fuel at an injection pressure of 1,850lb./sq. in. and is fitted with a vacuum-type governor operated by depression in the air intake manifold. The pintle-type injectors are recessed in the cylinder head, which latter is a one-piece iron casting incorporating the four combustion chambers and carrying the valves. The cylinder block and crankcase are a monobloc casting and replaceable wet liners are fitted, which will be appreciated from the economical maintenance point of view.

Forced-feed lubrication is used; a submerged pump in the sump draws oil through a filter and passes it to the oil galleries, and a pressure relief valve is fitted in an accessible position, so that oil pressure can be adjusted when necessary. The capacity of the system is 11.8 pints. Cooling is effectively done. The combined fan and circulating water pump are mounted on the front of the cylinder head and driven by a "V" belt from the crankshaft.

Starting is by a 24-volt electric motor assisted by the modern form of pre-heater plugs, and a 24-volt, 350-watt dynamo, driven by the fan belt, charges the battery. There are four forward speeds of 1.1, 1.6, 2 and 4.5 miles per hour, and in these the respective pulls are 4,500, 4,150, 3,750 and 1,725lb. The reverse speed is 1.1 miles per hour. It should be emphasised that all the traditional features that contributed so largely to the success of the original "Fiat 25 CI" tractor, particularly the high power-to-weight ratio and extreme manœuvrability, have been retained in the new model, whose efficiency is enhanced by the greater power, reliability and economy of the diesel prime-mover.

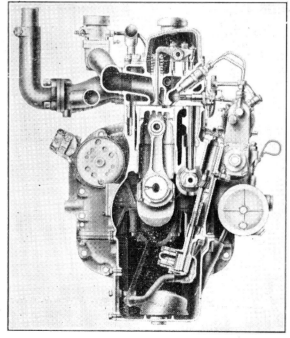

A Sectionalised View of the New Prime-Mover

Crawler Tractor Scrapbook Pt 2 — Fiat

Fiat 25C-25CS-25CI 1951

Fig. 5. - Tractor controls.

1. Engine water thermometer - 2. Magneto grounding push button, for stopping the engine - 3. Nearside steering clutch control lever - 4. Oil bath air cleaner - 5. Dynamo for lighting system - 6. Main clutch control lever - 7. Carburetor throttle control lever - 8. Pedals for independent action brakes on steering clutches - 9. Gear shifting lever - 10. Hand lever for tractor stopping brakes - 11. Offside steering clutch control lever - 12. Engine speed governor lever - 13. Fuel cock - 14. Radiator shutter control lever - 15. Auxiliary fuel tank filler cap - 16. Engine oil pressure gauge - 17. Instrument panel light.

Crawler Tractor Scrapbook Pt 2 — Fiat

Farm Implement & Machinery Review Sept.1961

Latest Italian Crawler Tractor
The Fiat "70.CI"

A 70 h.p. crawler tractor, intended for agricultural and industrial use, has been produced by the Fiat Company, Torino, Italy, and is being sold in this country by Mackay Industrial Equipment, Ltd., Feltham, Middlesex. The tractor, designated the "70.CI" is similar to the Fiat "60.CI," but several recent developments have been incorporated in the design.

The length of track on the ground has been increased and the track width is slightly greater. Power steering has been adopted, the controls of the multiple disc steering clutches and brake bands being hydraulically assisted.

A 4-cylinder, 4-stroke direct-injection diesel engine provides the power, and the gearbox is designed to give five forward and four reverse speeds. A "quick-reverse" unit, operated by a separate lever, links corresponding forward and reverse gear trains for rapid reversing during such operations as bulldozing.

Alternative starting methods are offered. A twin-cylinder 10½ h.p. petrol engine with a 12-volt electric starter can be fitted, or a 24-volt starter motor engaging directly with the tractor engine flywheel. There is a swinging drawbar with adjustable settings, and mud shields are provided above and below the track frames.

The price of the tractor, complete with bulldozing equipment, is £3,990.

FARM MECHANIZATION & BUILDINGS DECEMBER 1968

Fiats new farm line as advertised in UK in 1968 (without the wheel tractors).

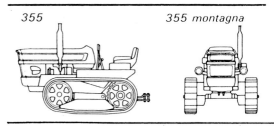

FIAT trattori *the largest wheeled and crawler tractor manufacturer in continental Europe*

Crawler Tractor Scrapbook Pt 2   Fiat

The Fiat 605CM was built in 1977. It has a three cylinder engine giving 60hp. This machine belonged to Eric Littlehales, Tokoroa, NZ in 1998 and its serial number was 1-807405.

Above: Fiat 555 belonged to Bill Higgs, Nelson when photographed in 1999.
Below: A Fiat publicity shot of a Fiat 100C.

Fiat Crawlers of 1978

Fiat 455 C - 48 CV

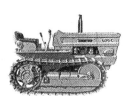
Fiat 505 C - 54 CV

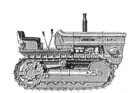
Fiat 605 C super - 66 CV

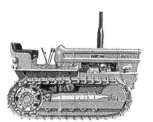
Fiat 805 C - 80 CV

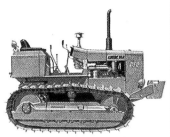
Fiat 90 C - 98 CV

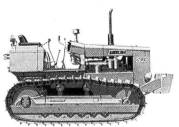
Fiat 120 C - 120 CV

Crawler Tractor Scrapbook Pt 2 — Fiat

The Fiat 65 Series was first produced in 1984 when two models were offered, the 70-65 and the 80-65. The 70-65 had a Fiat 8045.06, 4 cylinder, diesel engine.
The Fiat 80-65 pictured at the NZ National Fieldays was introduced to the NZ market in 1985.

The Fiat 75 Series was introduced to New Zealand in 1990. There were three basic models, the 60-75, the 70-75 and the 80-75.

The 60-75 had a Fiat 8035.05, three cylinder, diesel engine with a bore and stroke of 104 x 115mm. The 70-75 had a Fiat 8045.06, four cylinder, diesel engine with a bore and stroke of 100 x 115mm. The 80-75 the same engine as the 70-75 but with a bore and stroke of 104 x 115mm. The gearbox versions available were 8 forward and 8 reverse, 8 forward and 8 reverse and 16 forward and 8 reverse.

A Fiat publicity photo.

The Fiat 85 Series was first produced in 1993 with a 60-85, a 72-85 and an 82-85.

Fiat 72-85 had a Fiat 8045.06, four cylinder diesel engine with a bore of 100mm and stroke of 115mm giving 70 flywheel hp. It featured Steering-O-Matic "Full Drive". Weight with frame and roof was 3,600kg. Length was 3,240mm, height to frame was 2,220mm and width 1,650mm.
This machine was at the NZ National Fieldays in 1994.

Crawler Tractor Scrapbook Pt 2 — Fiat

FIAT Crawlers

Model	Production Start - Finish	Comments
AG4	1990s?	?
4C Mark III	1939 - 47	41.5hp, 4 cyl. petrol
4C Mark IV	1939 - 47	41.5hp, 4 cyl. petrol.
AD7, AT7C, BT7C	1971?	59.4dbhp, 4cyl. Fiat 604031 diesel
AD10, BD10	197?	102ehp, 79dbhp, 6 cyl. Fiat-OM CP3/100 diesel, AD10 Convert.
AD12, BD12	?	125ehp, 6 cyl.
AD14, BD14	1970-	150hp, 6 cyl., Fiat 20A diesel. AD/BD14 Convert.
AD18	1968-	?
AD20	?	?
25C /CS /CI	1951 - 53	23hp-kerosene, 28hp-petrol, 4 cyl.
25CD /CSD	1953 - 56	25hp, diesel 4 cyl.
25CDP /CIDP	1953 - 55	24hp, 4 cyl. petrol
25C /CSMk IV	1956 - 58	25hp, 4 cyl. diesel.
25C/CS Mk V	1956 - 58	24hp, 4 cyl. petrol
25C /CI /CF	1956 - 57	27hp, 4 cyl. OM COD/40 diesel.
OM30-40C	1952 - 55	40hp, 4 cyl, diesel
OM35-40CL	1954 - 55	35hp, 4 cyl. diesel
35	1952 - 60	35belthp, 29.7dbhp, 4 cyl.
40 Boghetto	1939 - 47	41.5hp, 4cyl. diesel
40CI	1960 - 67	40hp, 4 cyl. diesel
OM 45C	1956 - 58	45hp, 4cyl. diesel
OM45CI	1957 - 59	45hp, 4 cyl. diesel.
50 /50L	1946 - 50	50hp, 4cyl., diesel
OM50C	1959 - 67	50hp, 4 cyl. diesel
OM 50CI/CA	1960 - 69	49.9 belthp, 39.4dbhp, 4 cyl. diesel
50CI, AD5	1970 -	50hp, 38dbhp, 4cyl., diesel
52	1947 - 50	50hp, 38dbhp, 4cyl., diesel, track width only difference with 50
55	1950 - 53	55hp, 47.2dbhp, 4cyl., diesel
55-85	1995?	55hp, 3 cyl. diesel. 55-85V
60C	1956 - 60	60ehp, 50.25dbhp, 4 cyl. Fiat 604-010 diesel.
60CI	1956 - 61	60ehp, 4 cyl. Fiat 604-010 diesel
60CA	1956 - 58	60ehp, 4 cyl. Fiat 604-010 diesel
OM60CA	1962 - 63	60hp, 4 cyl, diesel
60-75	1990?	60hp, 3 cyl. diesel
60-85	1995?	60hp, 3 cyl. diesel. 60-85F
62-85	1993	60hp, 3 cyl.. diesel. 62-85M
70C	1961 - 70	74hp, 58.3dbhp, 4 cyl. Fiat 604-035 diesel
70-65	1984 -	70hp 4cyl. Fiat 8055.05
70-75CM SOM	1990?	70ehp, 4 cyl. diesel
72-85	1993 -	70hp, 4 cyl. diesel
80C	1971 - 75	80hp, 4 cyl., diesel
80-65	1984 -	80hp, 4 cyl. diesel
80-75	1990?	80ehp, 4 cyl. diesel
82-85M	1993 -	80hp, 4 cyl. diesel
88-85M	1993?	85hp, 4cyl. turbo. diesel
90C	1975 - 80	90hp, 4 cyl. diesel
95-55	1984 -	90ehp, 5cyl. Fiat 88055.05 diesel
100C	1971 - 75	100hp, 6 cyl. Fiat/OM CP3 diesel.
100-55	1995?	100hp, 6cyl. diesel
120C	1975 - 81	120hp, 6cvl. diesel
130C	1967 - 69	125ehp, 6cyl. Fiat 620-005 diesel.
150C	1970 - 74	160hp, 6cyl. diesel
160-55	1987?	160 hp, 6 cyl. turbo. diesel
180C	1966 - 71	180hp, 6 cyl, diesel
180-55	1988 -	180hp, 6 cyl. turbo diesel, prototype in UK in 1987, marketed ?
220C	1968	?

Crawler Tractor Scrapbook Pt 2 — Fiat

Model	Years	Specs
311C, 331, 351C	1959 - 61	27belthp 4cyl. diesel.
322C	1967	?
331CI	1960 - 61	27hp, 4 cyl. diesel
355C,	1971 - 75	35hp, 4 cyl. diesel, 355C Montagna
411C,	1958 - 68	45hp, 29.38dbhp, 4 cyl., diesel, Fiat 615.010
451C	1958 - 67	45hp, 4 cyl. diesel Fait 615.010
455CA/CI, AD4/M	1968 - 70	45hp, 3 cyl. Fiat 853 diesel.
455C/Montagna	1970 - 82	45hp, 3 cyl. diesel Fiat 853, also 455C Compatto
465CS	1982 - 84	54hp, 3 cyl. diesel
505C	1970 - 82	54hp, 3 cyl., Fiat 8035 diesel. 505C Montagna, 505C Vigneto
555C	1968 - 70	52ehp, 4 cyl. Fiat 845 diesel.
565C/M	1982 - 84	58hp, 3 cyl. diesel
601/601V	1949 - 50	petrol 21 belt hp, kerosene 17bhp, 4cyl.
605C/Montagna	1970 - 82	60hp 3 cyl. Fiat 8045 diesel,
605CM Super	1982 -	66hp, 4 cyl. diesel
655C	1971 - 75	65hp, 3 cyl. diesel
665C/M	1982 - 84	68ehp, 4 cyl. Fiat 8045 diesel.
700C (Type 30)	1932 - 40	30hp, 4cyl., petrol, **first Fiat crawler**
708 C (Type 20)	1934 - 43	20hp, 4 cyl. petrol
765C	1982 - 84	78ehp, 4 cyl. Fiat 8045 diesel.
805C	1975 - 80	80hp, 4 cyl., diesel
855C	1981 - 84 -	85hp, 4 cyl. diesel
955C	1981 - 84	90ehp, 5 cyl. Fiat 8055-04 diesel.
1355C	1981 -	135hp, 6 cyl. diesel

E & OE

TROUBLE WITH A FIAT 55 L

I cannot keep the gaskets oil-tight on the reduction gear casings of my Fiat 55 L. The trouble seems to be due to a build up of pressure after a few hours' work. Can this be obviated by fitting release valves to the reduction gear casings? If so, at what pressure should the valves open?

From our experience of isolated cases giving the same symptoms, it is more probable that the leakage is caused by slight movement of the casings due to wear in the bolt holes rather than from internal pressure which is not likely to exist anyway. To cure this movement three of the securing bolts should be substituted by bolts which have been specially machined on the plain or shanked portion so that they are a tap-in fit. These bolts will act as dowels, and to ensure that they can be tightened sufficiently, the length of the machined portion should be about 1/32-in. less than the thickness of the casing flanges.

Farm Mechanization December 1955

Below A new Fiat transporter for tracklaying tractors. The drive to the front wheels of the transporter is through the p.t.-o. shaft of the tractor.

Farm Mechanization November 1952

FOWLER, MARSHALL & TRACK MARSHALL

From PAST to PRESENT

The Fowler

John Fowler and Co. (Leeds) Ltd., pioneers of steam ploughing, are now specialists in Diesel tracklayer production

First Tracklayer

Fowlers built their first fully-tracked machine in 1923. It was a steam road locomotive with Fowler Snaketracs. But so far as the agricultural tracklayers were concerned, 1927 was the significant year because it was then that they introduced the Gyrotiller.

This was a giant machine which weighed 23 tons and was powered by a 220-h.p. engine using 14 gallons of petrol per hour. It stirred the soil to a depth of about 18 in. by six to eight vertical tines mounted on two discs which were carried side-by-side and rotated in opposite directions.

The Gyrotiller was originally designed for cultivating sugar plantations and the first model was put to work in Cuba in 1928. After the first four had been produced, the petrol engine was superseded by a 137-h.p. Diesel.

In 1932, the first of the Gyrotillers to be used in Britain was bought by an Essex farmer. Agricultural contractors then began to use them and thousands of acres on British farms were gyrotilled until about 1939, when production ceased.

In the meantime, Fowlers had experimented with a more conventional agricultural tracklayer and in 1934 they introduced the Fowler 3-30, which was powered by a 34-b.h.p. Diesel. By 1939 they had introduced three additional models: the 45-b.h.p. 4-40, the 70-b.h.p. 10-70 and the 86-b.h.p. Model 80.

These tractors marked the end of the steam age in British agriculture and in 1936 Fowlers made their last set of steam ploughing tackle; two years later they made their last steam engine of any type. Production of the new Fowler tracklayers was halted in 1939, when the factory was switched to the manufacture of war materials.

A Giant Gyro-Tiller Weighing 24 Tons at work in South Canterbury, NZ.

The FD Range

In 1945, Fowlers produced a new design of agricultural tracklayer in two models, the 28-b.h.p. FD-2 and the 35-b.h.p. FD-3. These remained in production until 1948, when Fowlers joined the Marshall Organization. The FD-2 and the FD-3 were then superseded by the 40-b.h.p. Fowler VF (now the VFA).

The current Challenger range of Diesel tracklayers began in 1950 when the Challenger 3 was introduced. It is powered by a 95-b.h.p. Diesel and has mounting points for earthmoving equipment. Steering is by clutch and brakes.

Early in 1952, the 50-b.h.p. Challenger 1 was put on the market. It was powered by a vertical, two-cylinder, forced-induction two-stroke engine and remained in production until late 1955.

The Challenger 4 was introduced at the beginning of 1953. It is the largest of the range and has a 150-b.h.p. six-cylinder Diesel. The most recent of the present range of Fowler Challengers, the Challenger 2, introduced last year, is powered by a 65-b.h.p. Diesel.

The Challengers are made in Fowlers' factory at Leeds, but the VFA is now being made by Marshalls at Gainsborough, together with the Track-Marshall, Field-Marshall and the Marshall M.P.6.

The 150-b.h.p. Diesel-engined Challenger 4 is the largest Fowler tracklayer.

THE FOWLER MARK VF

Drawing specially prepared for BRITISH FARM MECHANIZATION

KEY TO NUMBERS ON DIAGRAM.

1. 40 h.p. single cylinder 2-stroke Diesel engine. 2. Fuel injector. 3. Compression release valve. 4. Hand-start valve. 5. Ignition paper holder. 6. Cartridge starter. 7. Radiator. 8. Oil filter. 9. Oil filler cap. 10. Front drawbar. 11. Fuel filter. 12. Air cleaner. 13. Clutch and belt pulley. 14. Clutch brake pad. 15. Clutch operating fork. 16. First motion shaft. 17. Second motion shaft. 18. Power take-off—(b) power take-off control lever. 19. 3-speed change gear lever. 20. High-low change gear lever. 21. Governor control. 22. Clutch hand control. 23. Clutch foot control. 24. Steerage unit. 25. L.H. steerage band. 26. Tool box. 27. Final drive. 28. Track recoil spring. 29. Track tension adjustment. 30. Oil pipe to L.H. main crankshaft bearing. 31. Engine and transmission undershield. 32. Torsional front axle. 33. Rear cross beam. 34. Steering levers. 35. Swinging drawbar.

ENGINE SPECIFICATION

Single cylinder, horizontal, water-cooled, valveless, two-stroke Diesel.

Bore	6.5 ins. ; stroke, 9 ins.
Swept volume	298.65 cubic ins.
Compression ratio	16 : 1.
Belt horsepower	40 at 750 r.p.m. (max.).
Governor	Centrifugal, variable between 500 and 750 r.p.m.
Main bearings	Roller.
Big-end bearing	Upper half lead-bronze shell; lower half direct on to lead-bronze connecting-rod cap.
Crankpin diameter	3.5 ins.
Piston	Cast iron.
Gudgeon-pin bearing	Needle rollers.
Fuel	Gas oil.
Fuel injector pump and nozzle	C.A.V.
Injection pressure	2,000 lb. per sq. in.

GENERAL SPECIFICATION

Weight : 9,200 lb.
Overall length : 100.5 ins.
Overall width : 72 ins.
Overall height : 80.5 ins.
Ground clearance : 10.75 ins.
Turning radius : 114 ins.
Drawbar height : 13.5 ins.
Drawbar lateral movement, 26 ins.
Track gauge : 56 ins.
Shoe width : 14 ins.
Ground contact (each track) : 60.25 ins.
Total area of ground contact : 1,687 sq. in.
Ground pressure : 5.5 lb. per sq. in.
Belt Pulley (Standard Equipment)
 Dia. : 15 ins.
 Face : 6.5 ins.
 R.p.m. : 750 down to 500.
 Belt speed : 2,950 down to 1,970 ft. per min.
Position : Near side, centre.
Direction of rotation : Anti-clock, looking on pulley.
Capacities, Imperial gallons—five of which equal six U.S. gallons
 Radiator : 11 gallons.
 Fuel tank : 12 gallons.
 Oil sump : 7 pints.
 Transmission : 4.5 gallons.
 Final drive casing : 4.75 gallons (each side).
Extra Equipment
 Power take-off, side
 Power take-off, central driven from side p.t-o.
 Electric lighting
 Street plates
 Waterproof cover

Crawler Tractor Scrapbook Pt 2 — Fowler & Marshall

After much study of illustrations it has been decided that this is a Fowler 4/40. It was built in 1937 and had a Fowler Sanders four cylinder, diesel engine. Starting was by handle using a decompression device. It was at the Gore (New Zealand) rally 1997

Above:
This Fowler was photographed at the back of the local museum in Gunnedah, NSW, Australia in 1999. serial # 21785 engine # 688 4B makes it a Fowler 4/40. Beside it was Fowler serial # 22346 engine # M890 type 4B. There appears to be an Australian adaptation with the grousers on the tracks.

Left: The right hand side of the Fowler 4/40.

Crawler Tractor Scrapbook Pt 2 — Fowler & Marshall

Left: The Fowler FD3 had a three cylinder engine giving 35hp. It was made from 1945 to 1948. This restored model was photographed at an English rally by Richard Trevarthen.

Below: Another Fowler FD3 as found by Stuart Johnson, in England. It had been standing for more than a couple of years.

In a retirement village for old tractors in Western Australia, two Fowler VF crawlers enjoy the sun. The later VF with the bench seat is in front and behind is the earlier bucket seat type.

April, 1949 — BRITISH FARM MECHANIZATION

TRACTOR SERVICE

No. 1. The Fowler Mark VF

The Fowler Mark VF is one of Britain's outstanding tracklayers. Distinctive features are: single-cylinder Diesel engine, controlled transmission steering and track stabilizer. The track stabilizer replaces the more orthodox type of transverse spring and permits independent track frame oscillation over rough ground.

THE Fowler Mark VF is a full tracklayer powered by a single-cylinder Diesel engine. It has six forward and two reverse speeds. It weighs 9,200 lb. with belt pulley. A product of the Marshall Organization, it is made by John Fowler and Co. (Leeds) Ltd., England.

The engine is a counterpart of the well-known Field Marshall engine.

Starting

Engine starting is by hand or power cartridge. In the interests of economy, it is recommended that the power cartridge be used only when the engine is stiff as a result of extreme cold. As a safety measure use only cartridges bearing the " Marshall " name and trade mark.

To start by hand, disengage the clutch by pushing hand control lever forward (this must be returned immediately the engine starts). Remove ignition paper holder. Put the fuel control lever about two-thirds open and turn the engine two or three times (a sharp "purr" should be heard as fuel injection occurs). Put the cylinder-head valve in the hand-start position and set the decompression control roller in the outer of the threads turned on the periphery of the flywheel. Roll and insert ignition paper into holder. Light it, blow it until it glows and screw holder tightly into cylinder head. To prevent blow-by past holder, tap it lightly with a hammer. Turn engine by means of starting handle. The decompression roller will "unscrew" from the flywheel and cause compression. The engine should then start.

Always use ignition paper when hand starting, otherwise engine may back-fire.

To start by power cartridge, vary the hand-start settings by putting the cylinder-head valve in the power-start position and the decompression roller on the *inner* thread on the flywheel. In addition, remove the breech from the body, see that the firing pin does not project from the breech on the inside; remove protective end cap—if fitted—from cartridge, insert cartridge in breech body, and screw home breech cap. Insert the ignition paper as for hand starting and hit firing pin smartly—not heavily—with a hammer.

Immediately the engine has started, remove the cartridge case and set the cylinder head valve in the hand-start position. If the engine begins to "fade," pull the decompression-valve cable for a few seconds and then release it. The engine should then "pick-up." This applies also when hand starting.

Ignition paper should be used for power starting unless a start is being made within an hour of the previous run. If a cartridge does not fire, nothing should be touched for five minutes. Then extract the cartridge by completely unscrewing the breech cap and turning the engine slowly by hand. Do not attempt to prise out the cartridge, as it may detonate even after removal.

To stop the engine, run it slowly for about half a minute to allow fresh oil to circulate, and then move the fuel control lever to its foremost position.

Operation

Despite the fact that the engine is governed, it is possible to overload it. Overload will be indicated either by dark exhaust smoke or loss of engine speed. (Should these symptoms occur when a light load is being pulled, they will indicate engine out of condition due to broken air valves or dirty filter, etc.) With six forward and two reverse speeds available, continuous overload is inexcusable. When selecting a gear, remember that, provided the work will allow, it is always better to operate in a higher gear and reduce the travelling speed to the desired rate by reducing engine speed rather than operate in a lower gear with the engine at full speed. Speeds and maximum sustained drawbar pulls are as follows:

	m.p.h.	lb.
1st	1.26	10,000
2nd	1.69	7,800
3rd	2.34	5,300
4th	2.92	4,200
5th	3.9	2,700
6th	5.42	1,600
1st reverse	0.96	—
2nd	2.24	—

Engine Lubrication

A good-quality viscosity SAE40 straight mineral oil blended for compression-ignition engines is recommended for winter and summer. There is no objection to the use

(Continued overleaf)

As can be seen from this off-side-front view of the engine with bonnet removed, a high degree of accessibility, making for ease of decarbonizing, is a feature of the Marshall engine.

(A) Power/hand-start valve control; (B) decompression valve control cable; (C) exhaust drain elbow; (D) injection nozzle; (E) power-start pressure pipe.

Tractor Service: The Fowler Mark VF (contd.)

of a detergent oil provided the change is made strictly according to the oil manufacturer's recommendations.

The oil is fed to various parts of the engine through pipes extending from a multiple plunger pump located in the oil sump formed in the main bearing housing behind the flywheel. Surplus oil drains into the crankcase, whence it is forced by crankcase compression through a non-return valve and filter, from which it gravitates into the oil sump, the level of which should be checked twice daily.

Although the oil pump is a pressure unit, the lubrication system is not high-pressure in the sense that a full flush of oil is delivered to each lubricated point. Actually each plunger delivers a measured amount to its outlet feed pipe. The amount delivered to each pipe is controlled by a screw located above each plunger. These are correctly adjusted at the factory and should not be altered.

A quick approximate test of the oil pump can be made by turning the pump priming handle after the feed pipe (connected to the crankcase between the air cleaner and clutch adjuster) has been disconnected. The correct rate of discharge is a few drops per minute when the priming handle is being turned.

Air Cleaner and Valves

Air is drawn into the crankcase through a pre-cleaner, a main filter element and eight spring-steel air inlet valves, all of which comprise a single assembly attached to the crankcase. The pre-cleaner should be removed weekly (in extremely dusty conditions, daily) and shaken free of dust. On no account must it be oiled.

The main filter element should be examined at the same time: if it is dirty on the outside, the coconut-fibre filtering material should be removed, washed in paraffin, dried, moistened in warm lubricating oil, and repacked sufficiently tightly to prevent the passage of unfiltered air.

The spring-steel air valves are located between the filter and the crankcase. Their condition directly affects engine performance—if cracked, broken or gummed they will cause bad starting and loss of power. They should be examined every 250 hours and faulty valves renewed. A broken valve usually falls into the crankcase, whence it must be retrieved before the engine is started. When fitting a new valve, make sure it lies perfectly flat and fully covers the port opening.

Fuel Oil System

Two filters are incorporated in the fuel-oil system. The first is a screen fitted

Diagram of fuel and lubrication system of the Fowler Mark VF.

Should an increase in oil consumption occur, the most probable causes are: blocked filter element and/or sticking non-return valve.

The filter element should be removed and cleaned every 500 hours. When the element has been removed, the centre hole in the bottom of the container should be plugged to prevent dirt from entering while the interior is being cleaned. The element should be washed in clean paraffin, and renewed every 1,000 hours or when cleaning fails to reduce oil consumption.

The non-return valve is located in a casing secured by an oval flange on the underside centre of the crankcase. The inlet side is protected by a gauze. The valve comprises a ball retained by a centre screw. Dismantling and cleaning should be necessary only in the event of increased oil consumption which does not respond to either a clean or new oil-filter element.

To check the oil return system, remove the centre bonnet, start the engine and inspect the vent pipe. If oil is being blown from it, the system is faulty and should be checked as above. If puffs of air only can be felt, the system is in order.

Cooling System

No water pump is used, water circulation being entirely thermo siphon. The complete system can be drained through a tap located on the front near side of the engine. Cooling is by fan and radiator.

beneath the fuel tank filler cap. It should be removed only for periodic cleaning and not to facilitate tank filling. The second is an element assembly mounted to the rear of the air cleaner. The element should be removed and cleaned every 500 hours by forcing air through it from its centre while it is immersed in clean fuel oil or paraffin. To obtain the necessary air pressure, plug both ends of the element with corks, one of which contains a bicycle valve and pump air through the valve until the element becomes clean. Renew the element every 2,000 hours.

After the element has been disturbed, the fuel system should be air-bled in four stages: (1) pull the fuel control lever to its rearmost position; (2) open vent tap on the element container until air-free fuel flows; (3) slacken the bleeder valve located behind the plug below the inlet pipe to pump connection and retighten when air-free fuel flows; (4) remove ignition paper holder and turn the engine until the injection "purr" can be heard.

If a fully primed fuel pump fails to inject, examine the delivery-valve spring located under the pump outlet connection. (Do not turn the engine while the spring retainer connection is removed, or the valve will be ejected.) A broken or damaged spring should be renewed. Special precaution should be taken against the entry of dirt, the smallest particle of which can damage the nozzle.

Replacing a broken delivery-valve spring is the only on-the-spot fuel-pump repair recommended. Should other faults develop the pump should be exchanged for a service unit, available through distributors.

Fuel-injection timing is fixed at 6 degrees to 8 degrees before piston top dead centre. The pump is actuated by a cam integral with the drive gear and operating on the pump drive tappet. When fitting a fresh pump, it is important that this tappet be correctly adjusted by the removal or addition of shims between the tappet and pushrod. Adjustment can be checked by a mark through a window on the side of the pump. Shims should be added or removed until the mark stops within $1/32$ in. of the window end when the plunger reaches its maximum outward travel.

The fuel pump is controlled by a centrifugal governor mounted inside the oil sump. There are two methods of governor and pump adjustment: increasing the length of the rod connected to the hand-control lever decreases the engine speed, while increasing the rod length between the governor control spring and fuel pump decreases the fuel supply. Shortening these rods increases engine speed and fuel supply respectively. These adjustments should be made only by an expert equipped with a revolution counter. The maximum speed should not exceed 800 r.p.m. when the engine is running idle.

Injection Valve

The injection valve assembly comprises a nozzle valve and a pressure mechanism set to discharge at 2,000 lb.

Adjustment of the pressure mechanism demands specialized knowledge and equipment, but the nozzle valve unit screwed on to the discharge end of the assembly can be renewed on the spot. To field-test the assembly, withdraw it from the head, connect the injection pipe and, with the fuel-control lever in the open position, turn the engine. A clean discharge of fine spray in the form of a cone, accompanied by a sharp purr, will denote perfect condition. An irregular stream of drops will denote unsatisfactory condition, the most probable cause being dirt on the valve seat. A faulty valve should be unscrewed from the main body and washed in clean fuel. Obstinate carbon should be softened in fuel and rubbed away with nothing harder than a piece of wood.

Exhaust System

The Marshall engine is no exception to the rule that the condition of the exhaust system of a two-stroke engine has a greater bearing on engine efficiency than that of a four-stroke engine. The exhaust pipe should be dismantled and cleaned every 500 hours or more frequently if the engine is on light work continuously. If the system is not cleaned, it may become oil-soaked and catch fire. Should this occur, reduce engine speed and allow the fire to burn out.

The exhaust port in the cylinder wall should be cleaned each time the pipe is dismantled. To prevent carbon falling into the cylinder the flywheel should be turned until the piston fully covers the port.

Decarbonizing

Perhaps the most accurate indication that the piston should be withdrawn for decarbonizing and ring cleaning can be obtained by inspection when the exhaust port is being cleaned. If the piston and rings appear black, they should be removed and cleaned. This is a simple operation involving removal of head, air cleaner assembly and big-end nuts. Note should be taken of the position of the rings so that they can be replaced in the same order. When replacing the head,

April, 1949 — BRITISH FARM MECHANIZATION

tighten the nuts progressively in opposite pairs.

A characteristic of the Marshall engine is that the upper half of the big-end bearing is subject to a much greater percentage of the load than the lower half. Consequently, it is usually only necessary to renew the upper half in the event of normal wear. The correct running clearance of a new bearing is .006 in.—on no account must it be less. A .050 undersize bearing is available for a reground crankpin. The correct tension of the big-end bolts is 98 ft./lb.

The cylinder block, being a comparatively simple casting, is relatively inexpensive to make, consequently the provision of a replaceable cylinder liner as a means of reducing maintenance costs is not deemed necessary. It has proved cheaper to rebore or renew the block. The maximum diametric amount of cylinder wear permissible before either of these operations is performed is .018 in. when measured vertically at 1.25 in. from the top of the cylinder.

Engine Clutch

The engine clutch is a multi-fabric-disc-steel plate unit contained in the belt pulley and mounted on the near side of the crankshaft. Normal operation is controlled by a foot pedal. The first movement of the pedal disengages the clutch, and the second movement presses the back of the clutch centre against two brake pads attached to the crankcase. Adjustment for wear is by two nuts on the control cable eyebolt located on the crankcase above the clutch. These should be slackened sufficiently to give 0.5 in. free movement of the pedal. The brake pads should clear the back of the clutch by 0.125 in. when the clutch is fully engaged. This clearance should be maintained by the addition of shims supplied with the machine.

An auxiliary clutch control hand lever is for declutching when starting, etc. When the engine is running, this method of declutching should be used for very short periods only because of the strain imposed on the mechanism.

The clutch can be relined without disturbing adjacent components. The repair is a simple one which can be done by an expert mechanic in less than two hours. Unless the clutch has been persistently abused, the steel plates should outlast four to five sets of fabric discs.

"Balanced Power" Unit

From the clutch, engine torque is transmitted to the track drive sprockets by "all-spur" gears. An interesting feature is the patented "Balanced Power" differential and steering unit. This is splined on to the inner ends of the two half-shafts which carry the final-drive reduction gear pinions—one on each outer end. Each half-shaft extends through a hollow shaft.

The hollow shafts are the steering-brake tubes and a brake drum is mounted on the outer end of each. The inner ends are positively attached to the Balanced Power Unit. When a steering brake is applied, a redistribution of power is effected. The speeds of the braked shaft and half-shaft decrease, while the speeds of the opposite shafts increase, consequently the machine alters course. This type of steering ensures a gradual turn by preventing either track being locked.

Correct lubrication and brake adjustment are the essential maintenance features. One oil supply is common to the gearbox, "Balanced Power Unit" and brakes. A high-quality SAE 140 viscosity oil should be maintained to the level denoted on the dipstick, and changed every 2,000 to 3,000 hours. Each brake is adjusted by a stud, screwed into the back of the brake chamber. Correct adjustment permits 3½ in. free movement at the top of each steering lever.

The brake adjusting studs should not be confused with the brake-band support studs, one of which is located in the bottom of each brake chamber. If these support studs have been disturbed, they should be reset by being screwed inwards until fingertight, and then backed off 1½ turns.

Final-drive lubrication is independent of the transmission, but the same grade of oil should be used. Each housing should be drained every 2,000 to 3,000 hours.

Although detailed instruction on how to dismantle the transmission and final drives is beyond the scope of this article, it is interesting to note the high degree of accessibility achieved by the manufacturers: the gearbox shafts, bearings and pinions can be removed without disturbing affiliated components; the sprockets and the final-drive and brake assemblies can be removed without disturbing the track frames. Despite the simplicity of these operations, however, it is strongly recommended that they be done by an expert mechanic.

Tracks

The track-drive sprockets can be reversed by transference to opposite sides. The track chains comprise steel pins and bushes pressed into steel links. The bushes are interlocking—a type designed to exclude dirt from the bush interior. Link pitch is 6¾ ins. Steel track shoes integral with grousers are bolted to the chains.

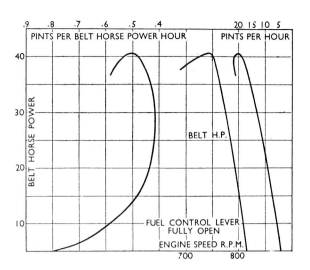

The graph shows engine performance curves.

Correct track tension is an essential of economic track maintenance. Adjustment is by a turnbuckle located behind each front idler. Correct tension is when each track can be raised approximately 2 in. from the top idler roller. Even more important than correct tension is the turning of pins and bushes. This should be done when the internal and external wear on the bushes at sprocket tooth point of contact has reduced the original dimensions by one-third.

When the pins and bushes are turned, the sprockets should be reversed. The bottom rollers, top idlers and front idlers should also be inspected. It is quite probable that the rear rollers will have worn more than the front. To counteract this, they should be exchanged with either the top idlers or the front bottom rollers. More detailed information on track maintenance will be found on pages 16, 17 and 18.

A unique feature of the VF is the track-frame stabilizer which carries the front of the tractor, maintains track-frame alignment, and permits independent vertical oscillation of the track units through a range of 6 ins.

The manufacturers are John Fowler and Co. (Leeds) Ltd., Leathley Road, Leeds, 10.

A sectional drawing and specifications of the Fowler Mark VF tractor appear on the following pages.

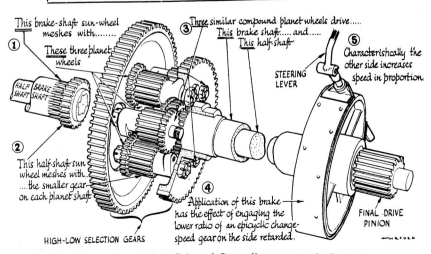

This illustration of the "Balanced Power" steerage unit, incorporating a two-speed rear axle assembly, shows how a spur-geared differential with compound planets and sun wheels, maintains a positive drive at reduced speed to the braked half of the assembly.

55

May, 1951 — FARM MECHANIZATION

Fig. 1.

The Fowler Challenger Mark III

Powered by a 95 b.h.p. 6-cylinder Diesel engine, the Fowler Challenger Mark III is a recent addition to the range of heavy British tracklayers in commercial production. It provides 82 drawbar horse-power and a bottom gear pull of 21,100-lb.

DESIGNED for bulldozer, scraper and heavy agricultural duties, the Fowler Challenger was extensively tested in prototype form before line production began towards the end of 1950 in the Leeds factory of John Fowler (Leeds), Ltd. Current production is at present mainly confined to the home market, but Fowlers have announced that shipments will soon be sent to Australia, Canada, French North Africa, France, Italy, Sweden, India, Pakistan, Brazil, Nigeria, Venezuela, Bolivia, Syria, Jordan, Norway, the Belgian Congo, Malaya and other overseas markets.

Priced at £3,525, the Challenger weighs approximately 23,600 lb. Its main features are a Meadows six-cylinder Diesel engine, electrical starting equipment, fabricated-steel hull and main transmission housing, clutch and brake steering, steel tracks with press-fitted steel pins and bushes, detachable plates, track frames incorporating Fowler patented articulated suspension, power drives for front-mounted hydraulic pump and rear-mounted winch. Provision is made for Blaw-Knox, Bray and Onion bulldozer and auxiliary earth-moving equipment.

Engine

The engine is a four-stroke, six-cylinder, direct-injection Diesel which is governed to a maximum of 1,550 r.p.m., at which speed it develops 95 b.h.p. A single casting forms the crankcase and cylinder block. The latter contains renewable dry liners. The pistons are aluminium alloy and each carries three compression and two scraper rings. Pistons and connecting rods can be removed from the top of the cylinders.

There are seven crankshaft main bearings. These and the big-end bearings are renewable, precision, steel-backed, lead-bronze shells. They are pressure-lubricated at approximately 50 lb. per sq. in. from a gear-type oil pump which also forces oil to the gudgeon pins, valve gear, camshaft and timing gears.

Two filters and a pressure relief valve which lifts at 50 lb. per sq. in. are incorporated in the lubricating system. The first filter is a floating screen which protects the pump inlet. It should require attention only when the engine receives a major overhaul. The second filter is a four-element full-flow assembly which is mounted on the underside of the sump and intercepts all the oil from the pump. The elements in this assembly should be cleaned every 200 hours and renewed every 400 hours.

Oil pressure is registered by a gauge on the instrument panel. The minimum safe pressure is 25 lb. per sq. in. Sump oil capacity is five gallons. S.A.E.20 is the recommended viscosity for ambient temperatures below 40 degrees F., S.A.E.30 between 40 and 80 degrees F., and S.A.E.40 above 80 degrees F.

Cylinder Head and Valves

There are two cylinder heads. These are interchangeable and carry the valves, injectors and manifolds. The valve seats are renewable inserts and the valve-tappet clearances, hot, are: inlet, 0.010 in.; exhaust, 0.012 in. Orthodox adjusters are provided and it will facilitate adjustment to remember that the firing order is 1, 5, 3, 6, 2, 4.

An inlet-valve decompressor is fitted as a starter aid. After the inlet-valve tappet clearances have been adjusted, the compressor clearances should be checked and adjusted to provide 0.015 in. between the flats on the decompressor shaft and the auxiliary tappets.

Clean Air and Fuel Essential

The air and fuel systems comprise twin oil bath, air cleaner, C.A.V. injection pump, four-hole injectors, filters and a 57-gallon fuel tank. Clean air and fuel are more critical to the Diesel engine than to any other type. Neglect in this sphere of operation is responsible for most Diesel engine stoppages. Correct maintenance routine is described in the driver's handbook, and the following comments are made to emphasize the need for observing these instructions.

When the engine is doing 1,500 r.p.m. it is receiving 4,500 fuel injections per minute. In other words, 270,000 separate units of fuel are sprayed into the engine in one hour. A consumption of, say, three gallons per hour would mean that each fuel unit was 1/90,000 of a gallon. The finish of the components that produce this precision is so fine that scoring invisible to the naked eye will destroy the accuracy required for efficient engine performance. Furthermore, these components can be damaged by abrasive particles also invisible to the eye. To ensure the removal of such minute particles from fuel in which they may be suspended, each consignment should be allowed to settle for at least a day after delivery.

Two storage tanks may be required to allow this. Each tank should slope for

Fig. 2.
Left-hand side of engine, showing breather (A), fuel oil filter (B), governor housing (C), dynamo (D), C.A.V. injection pump (E), and reduction gear-box bolted to flywheel housing (F).

The Fowler Challenger Mark III (contd.)

about ½ in. per foot of length. Fuel should be drawn from the higher end and sludge from the lower. If fuel has to be transported from storage tank to tractor, every precaution must be taken to overcome the risk of contamination. If possible, fill the fuel tank before leaving the tractor shut down and thus drive out vapour and reduce condensation.

The fuel tank of the Challenger is of heavy section welded steel. It is situated beneath the propeller shaft and gearbox and is a major component of the tractor hull. The air cleaner has two oil baths and cleans the air in three stages: large particles of dirt are driven into the first oil bath by centrifugal action; intermediate particles dive into the second oil bath, and the finest particles to survive these hazards find themselves trapped in an oil-soaked element through which the air must pass to leave the cleaner.

Two Filters in Fuel System

There is a filter in the fuel tank inlet, as well as two filters in the fuel system: the first is incorporated in the suction lift pump and the second between this pump and the injector pump. If these filters are found to be clean when dismantled at the intervals stated in the instruction book, it will indicate that clean fuel is being provided, in which case the intervals between filter cleaning can be lengthened with advantage.

The injectors are set to provide an injection pressure of 2,540 lb. per sq. in. There should be an audible *ping* at each injection when the engine is turned. If there is no such noise, clean the nozzle and spray holes. Should this fail to effect improvement, the injector should be serviced by an expert with specialized equipment.

Lubrication required by the fuel injection pump varies according to which of two types of pump is fitted. One incorporates a dipstick and oil filler cap, in which case oil should be kept to the level shown by the dipstick. The other type, recognizable by having no dipstick, is lubricated automatically.

A mechanical governor is incorporated in the pump and is interconnected with a hand control lever by which any speed between 500 and 1,550 r.p.m. can be predetermined. A Meadows' patent injector timing device in the pump drive automatically varies injector timing according to engine speed.

The cooling system is a pump-assisted thermosiphon and includes a belt-driven fan. The correct fan belt tension allows ¾-in. flex between pulley centres and is adjustable by a flange on the fan pulley.

Electrical equipment is 24-volt, and comprises a dynamo, 75 or 81 amp./hr.-capacity batteries and an axial type of starter motor. The maximum charging rate of 10 amps. is shown by an ammeter.

The engine clutch is a spring-loaded single plate, 16 ins. in diameter and foot operated. An adjuster for maintaining the correct free movement of the pedal is incorporated in the pedal linkage. From the clutch, engine torque is transmitted to a speed reduction box, containing spur gears, bolted to the engine flywheel housing. The removal of this box allows the engine clutch to be taken out. A high-grade S.A.E.140 oil should be used in the reduction box which has a capacity of 6 pints.

From the speed reduction box, a short propeller shaft transmits the drive to a gearbox bolted to the bevel gear and steering clutch housing. There are three gear selection controls: clutch pedal, normal gear lever and a secondary gear lever. When any of the four lower speeds has been selected by the normal gear lever, this speed can be changed to or from reverse by a forward or backward movement of the secondary lever after the clutch has been disengaged.

The gearbox can be removed without disturbing adjacent assemblies. It needs the same grade of oil as the speed reduction box and, in conjunction with the bevel gear compartment, has a capacity of 13 gallons. The bevel gear compartment is the centre of three compartments contained in the fabricated steel housing. Each of the two compartments flanking the bevel gear contains a steering clutch and brake assembly.

A single lever operates each clutch and brake and is arranged so that the clutch is fully disengaged before the brake is applied. Contracting band brakes are used and these are self-wrapping irrespective of whether the tractor is in forward or reverse gear. Both brakes are interconnected to a single pedal for parking. Correct steering lever free movement is 3 ins., while the pedal requires 2 ins. The brake bands can be removed for repairs from the top of the housing.

Fig. 3.—Three-quarter view of underneath of fabricated steel hull. The channels accommodate the shafts shown in Fig. 4.

Fig. 4.—Front view of track frames and suspension. The shafts support the weight of the tractor. The front shaft allows each frame to pivot on the rear shaft.

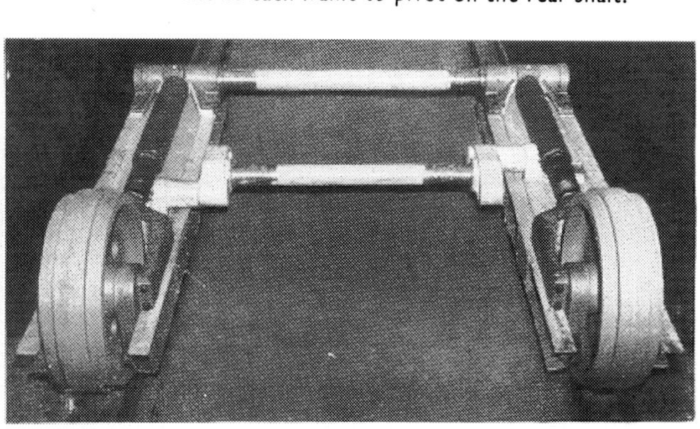

May, 1951   FARM MECHANIZATION

Fig. 5 (left): Final-drive reduction-gear housing showing sprocket hub and steering-clutch drum.

Fig. 6 (right): Bevel gear and steering clutches. A clutch can be removed after the reduction-gear housing (*Fig. 5*) has been detached.

To remove a steering clutch, it is first necessary to detach a final-drive housing. This can be done without affecting the adjacent track frame. Each final-drive housing contains a pair of straight-toothed reduction gears, the larger of which is mounted on the sprocket shaft.

Five bottom rollers, two top, and one front idler are attached to each channel steel track frame. The bottom and top rollers are interchangeable. They are carried on centre flanged axles and have split bronze-bushed hubs which are interchangeable with the front-idler hubs. The front idler is spring-loaded to absorb track shock recoil and also provides for track adjustment. The correct track tension allows 2 ins. slack at the track top when the rest of the track is tight. Each track bush is case-hardened to a depth of 0.1 in. and it is suggested that the pins and bushes be turned when the external diameter is reduced to $1\frac{9}{16}$ ins. The bushes are recessed into the outer track links.

The track frame suspension conforms to the design which has proved successful on the Fowler VF. A transverse shaft immediately in front of the sprockets supports the tractor on the rear of the track frames, and a double-cranked shaft supports the forward weight on the centre of each frame. The frames are free to pivot on the rear shaft, and the cranked shaft is free to turn. Thus, opposing vertical movement of the frames is obtained without sacrificing that lateral stability so essential to track life.

Swinging Drawbar

A swinging drawbar is anchored under the centre of the hull, which, apart from a slight profile made by this anchorage, presents a flush face likely to be of great value in the rough, rocky and muddy terrain in which many a Challenger will be expected to prove the qualities implied by its name.

ABRIDGED SPECIFICATION OF THE FOWLER CHALLENGER MARK III.

Engine.—Meadows Diesel, type 6 DC 630. Vertical six-cylinder, in-line, compression-ignition, cold-starting, four-stroke. Bore, $5\frac{1}{8}$ ins.; stroke, $5\frac{1}{4}$ ins.; capacity, 633 cubic ins. (10.35 litres). Compression ratio, 16 to 1. B.h.p., 95 at maximum governed speed of 1,550 r.p.m.

Monobloc crankcase and cylinder block.—Renewable dry liners. Seven main bearings crankshaft. Renewable, steel-backed lead-bronze main and big-end bearing shells.

Head and valves.—Two interchangeable cylinder heads with renewable valve seat inserts. Tappet clearances, hot, inlet, 0.010 in.; exhaust, 0.012 in.

Valve timing.—Inlet opens 18 degrees before top dead centre, closes 34 degrees after bottom dead centre. Exhaust opens 40 degrees before bottom dead centre, and closes 8 degrees after top dead centre.

Lubrication system.—Pressure feed from gear-type pump. Full-flow filters. Pressure release valve lifts at 50 lb. per sq. in.

Air, fuel and injection equipment.—Twin oil-bath air cleaner. C.A.V. type BPE 6B80Q pump incorporating mechanical governor. Four-hole injector nozzles. Injector pressure, 2,540 lb. per sq. in. Automatic injection advance and retard in pump drive.

Starting aids.—Excess fuel device on pump. Inlet valve decompressor.

Electrical equipment.—24-volt, comprising dynamo (10 amp. charging rate), 75- or 81-amp./hour-capacity battery (10-hour rating), axial type of starter motor.

Engine clutch.—Single plate, 16 ins. diameter, dry, foot-operated.

Speeds and drawbar pulls.—Six forward and four reverse.

	Forward (m.p.h.)	Pull (lb.)	Reverse (m.p.h.)
1st	1.50	21,100	2.08
2nd	2.25	13,200	3.10
3rd	3.00	9,250	4.15
4th	3.75	6,870	5.13
5th	4.37	5,530	—
6th	5.67	3,640	—

A twin-range gearbox giving 12 forward (maximum 10.65 m.p.h.) and 8 reverse speeds is available as extra equipment.

Steering.—Multi-plate clutches and brakes—controlled by two levers. Brakes interconnected to foot pedal for parking.

Final drive.—Straight-toothed reduction gears to cast steel track sprockets. Can be removed without disturbing track frames.

Tracks and track frames.—Centre-strutted steel links, press-fitted steel pins and bushes (latter recessed into outer links), bolt-on 20-in.-wide track shoes with integral grousers (18-in., 22-in. and 24-in. shoes also available). Bottom rollers of steel with split hub bronze bearings and centre-flanged axles. Rollers interchangeable with top idlers, hubs interchangeable with front idler. Vertical oscillation of track frame permitted by Fowler patent double crank, which supports front of tractor on track frames. Track adjustment via spring-loaded front idler. Correct tension allows 2 ins. slack.

Other Details

Weight	23,600 lb.
Track centre	72 ins.
Track plate width (standard)	20 ins.
Length of each track ground contact	94 ins.
Total area of ground contact	3,760 sq. ins.
Ground pressure	6.3 lb. per sq. in.
Overall length	$162\frac{1}{4}$ ins.
Height to top of exhaust	$106\frac{1}{2}$ ins.
Ground clearance	15 ins.

Capacities.—Radiator, 9 gallons; sump, 5 gallons; reduction gearbox, $\frac{3}{4}$ gallon; gearbox and bevel gear compartment, 13 gallons; final-reduction housings (each), 4 gallons.

Manufactured by John Fowler and Co. (Leeds) Ltd., Hunslet, Leeds, 10.

Crawler Tractor Scrapbook Pt 2 — Fowler & Marshall

The Fowler Challenger Mark III

(Powered by a 95 b.h.p. 6-cylinder Meadows Diesel engine)

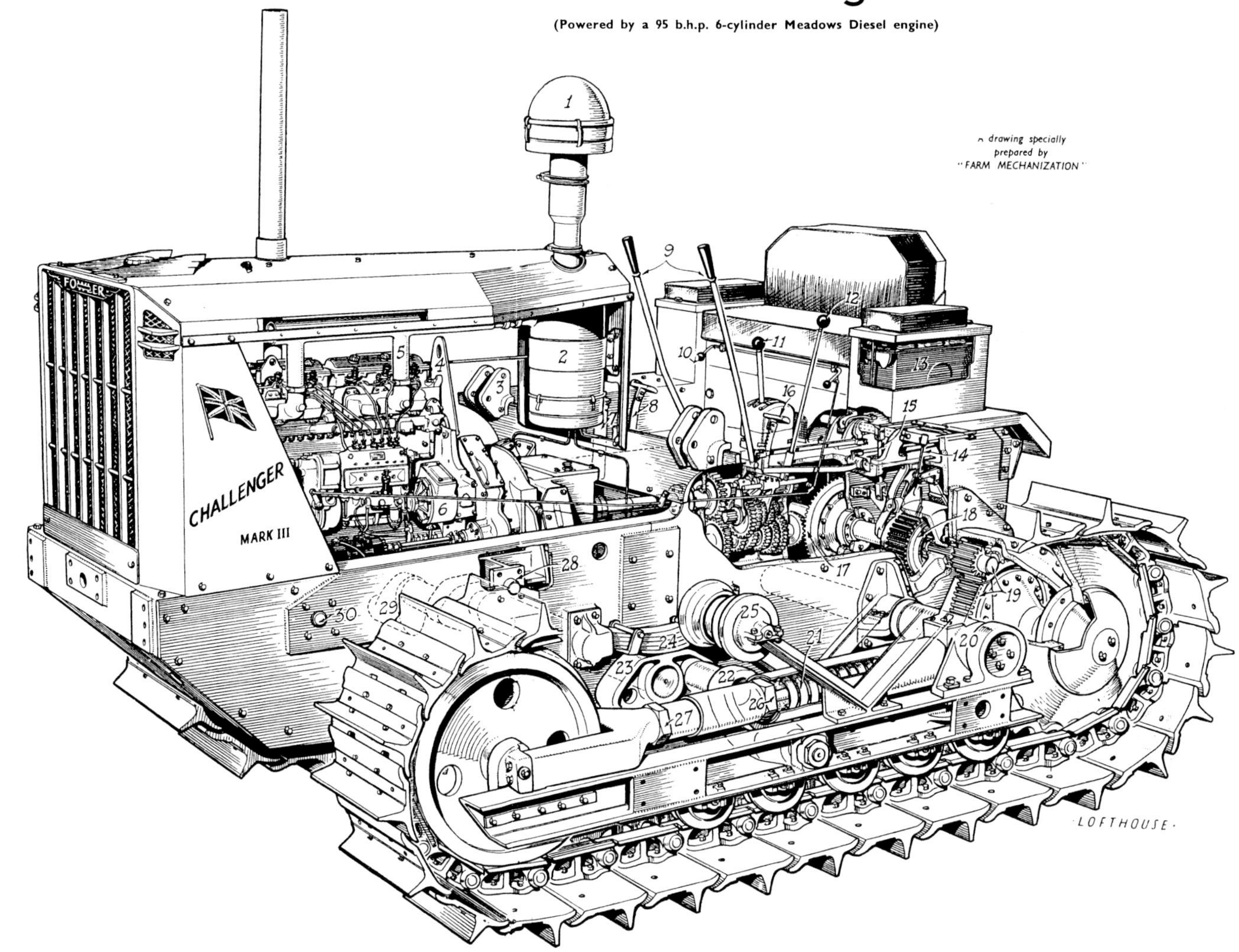

a drawing specially prepared by "FARM MECHANIZATION"

KEY TO NUMBERS ON DRAWING

1. Primary oil-bath air cleaner.
2. Main oil-bath air cleaner.
3. Bulldozer hydraulic ram bracket (R.H.). Can be used with L.H. bracket to hoist tractor.
4. Engine lift bracket, rear.
5. Inlet manifold.
6. Automatic variable-injection timing housing.
7. Fuel-tank filler.
8. Engine clutch pedal.
9. Steering clutch and brake levers.
10. Throttle.
11. Gear lever.
12. Forward and reverse lever.
13. Battery.
14. Steering-brake connecting rod.
15. Steering-clutch connecting rod.
16. Parking brake pedal.
17. Bevel pinion and crown wheel.
18. Left-hand steering clutch.
19. Final-drive reduction gears.
20. Track-frame pivot bracket, rear.
21. Track recoil spring.
22. Front-beam crank bracket (L.H.).
23. Cranked front beam.
24. Crank damper spring.
25. Top track carrier roller.
26. Track adjuster nut.
27. Track adjuster locknut.
28. Engine mounting, rear.
29. Front-beam crank bracket (R.H.).
30. Engine mounting, front.

WORN GEAR SELECTOR

I have a Fowler VF tractor which is just over two years' old and I am having trouble with it. When the H/L gear lever is in the L position, the main gear lever has a tendency to come out of the selector gate when I move out of 3rd gear. I have had the gearbox top cover off and the parts do not seem to be worn.

Secondly, after tightening the left-hand track the required amount I find that it has a tendency to toe-out at the front.

Although the selector mechanism may appear unworn, it is possible that sufficient wear has taken place to cause the condition you describe. When the two change-speed selector forks are in the neutral position, there should be $\frac{1}{16}$-in. clearance between the two forks. The faces against which the lever tip operates should be in good condition. If the clearance is excessive, either renew the forks or rebuild them by bronze welding.

It is not possible to advise specifically on the condition of the tracks without personal examination. We suggest that you check all the connections and bearings of the track frames and cross beams for excessive wear. If these parts are in good condition, the trouble may be due to incorrect adjustment.

July, 1950 — BRITISH FARM MECHANIZATION

New Fowler Tracklayer

A new Fowler tracklayer and an improved Field Marshall wheel tractor have been announced by the Marshall Organization.

The tracklayer, to be known as the Challenger Mark II, has been designed for industrial and heavy agricultural work. It is powered by a vertical, twin-cylinder, two-stroke Diesel engine developing 80 b.h.p. Steel-plate hull construction gives a flush underbelly and uniform ground clearance. It is equipped with clutch and brake steering.

The Field Marshall Series 3 wheel tractor is fitted with a twin-range gearbox giving six forward and two reverse speeds.

A double-cranked torsion beam on the Fowler Challenger Mark II (*above*) allows independent track frame oscillation over rough ground. Substantial mountings and drilled facings are integral parts of the basic tractor design and are provided to permit use of a full range of ancillary equipment.

THE Challenger Mark II tracklayer weighs approximately 23,600 lb., and has six forward and four reverse speeds. Drawbar pull in 1st speed, 1.5 m.p.h., is 17,600 lb., and 2,980 lb. in 6th speed of 5.67 m.p.h.

The engine is a Marshall vertical twin-cylinder two-stroke Diesel, which develops 80 b.h.p. at 825 r.p.m. Bore and stroke are $6\frac{1}{2}$ ins. and 8 ins. respectively. Starting is by twin cylinder petrol engine with friction clutch and Bendix pinion.

The engine clutch is a foot-operated, spring-loaded dry-plate disc with clutch brake. Steering is by dry multi-disc fabric-type clutches and contracting brake bands. Each brake and clutch is controlled by a single lever.

Suspension is by double-cranked torsion beam, which permits independent track-frame oscillation over rough ground.

Track Details

The tracks are pin-and-bush type with steel plates and grousers. There are five bottom rollers to each track frame. The frames are of fabricated steel channel hinged on a pivot beam mounted to the main frame. There are two top-carrier rollers to each track. These are interchangeable with the bottom rollers. Track adjustment is via the front idler wheels, and track recoil springs are incorporated in each frame.

Three power take-off points are provided, one at the rear and two at the front.

General dimensions are: track centre distance, 72 ins.; track plate width, 18 ins.; track ground contact, each track 94 ins.; ground pressure 7 lb. p.s.i.; overall length 13 ft. $6\frac{1}{4}$ ins.; overall width, 7 ft. 6 ins.; ground clearance, 15 ins.

The Challenger Mark II tracklayer is manufactured by John Fowler and Co. (Leeds) Ltd., Leeds, 10.

INJECTOR PUMP TIMING

Please tell me how to time the fuel injector pump to my Fowler FD3 engine.

Each fuel pump plunger must be timed to the cylinder it supplies. No. 1 piston is that nearest to the flywheel and No. 1 plunger is on the left when the pump is viewed from the pump side of the engine. Install the pump on its support and connect the main feed pipe but do not replace the fuel delivery pipes. Remove the delivery valve holder from No. 1 discharge, lift out the valve and spring and replace the delivery valve holder. Rotate the flywheel in its normal direction until its No. 1 dead centre mark is in line with the timing pointer, and the piston is at the top of its compression stroke—with both valves closed.

Turn on the fuel tap and set the engine speed control in its wide open position. Rotate the flywheel backwards very, very slowly, to a position which allows fuel to flow in a continuous stream from No. 1 delivery valve holder. From this position, rotate the flywheel in its normal direction, again very slowly—and while maintaining a supply of fuel by hand operation of the lift pump—and observe the spill of fuel from the valve holder. At a certain point, the spill will stop. It is most important to identify this point precisely. It indicates the start of fuel injection and should coincide with the appearance of No. 1 INJ on the flywheel in register with the timing pointer. *Note that this position should always be approached from the normal direction of the flywheel rotation.*

If the cut-off occurs before the flywheel mark reaches the pointer, the timing will be too far advanced. To retard this, remove one or more shims from beneath the pump tappet cap. If the timing is retarded, it will be necessary to add shims at this point. A shim of 0.002 in. is equivalent to approximately $\frac{1}{4}$ in. movement of the rim of the flywheel. Check each cylinder against the appropriate marking on the flywheel and adjust the shims when necessary.

It is most important, during the foregoing operation, that the fuel supply is maintained by actuation of the lift pump.

Crawler Tractor Scrapbook Pt 2 — Fowler & Marshall

Right: The earlier Fowler VF with the single seat is very difficult to find in original condition in New Zealand. The large majority had the seat converted to something more substantial.

This Fowler VF was photographed at a rally in Western Australia in 1999 by Harald White.

Below: This Fowler VF is still used by its owner, Bruce Gore, Dannevirke, New Zealand, on his small holding. It was photographed in 1998. The Fowler Mark VF had a two stroke, single cylinder engine with a 6½inch bore and 9 inch stroke giving 40hp.

Involved in a "slow race" at Ashburton, New Zealand, in 1990 it looks as if the Allis Chalmers D21 has the slower gear ratio than the Fowler VF. The VF had six forward speeds and two reverse, low 1.26, 1.69 and 2.34mph, high 2.92, 3.8 and 5.42mph. Reverse speeds were .96 and 2.24mph.

Crawler Tractor Scrapbook Pt 2 — Fowler & Marshall

Don Heslop, Nelson NZ, shows off his Fowler VFA in 1999. The VFA was powered by a Marshall Series 3A, two stroke single cylinder engine giving maximum belt hp of 40.3, drawbar hp was 34.3. The engine had a 6½ inch bore and 9 inch stroke. It had six forward speeds and two reverse. Total weight was 10,510lbs, length was 8ft 4½ inches, width was 6ft and height was 6ft 10 inches. It was made from 1952 to 1957.

This early model Fowler VF crawler is a bit of a mystery. It was made before 1950, some six years before the Track Marshall appeared but has Track Marshall written on it in several places. The seat is not original. The tractor was photographed at the Whangarei District Museum in 1998

Introducing the "Challenger 4"

Another Powerful Addition to the "Fowler" Range of Diesel Crawlers

Another highly-impressive addition to the existing range of diesel crawler tractors designed and produced by Messrs. John Fowler & Co. (Leeds), Ltd., Hunslet, Leeds, 10, emerges with the introduction of their new "Challenger 4" machine. Subjected to extensive testing over a period of more than two years, it is the heaviest of all "Fowler" models, and, apart from its great potential value in large-scale development schemes for agricultural purposes, should soon engage the interest of those concerned with expanding civil, industrial and military undertakings as well. The power harnessed to meet the most exacting of requirements is considerable.

Merely to list the main features of the "Challenger 4" tractor, without any elaborating detail, should provide readers with a good idea of its operational possibilities. In addition to the "Meadows" 6-cylinder diesel engine with auxiliary petrol engine starting, the fabricated steel hull and main transmission housing, and clutch and brake steering, there are hardened steel tracks with press fitted steel pins and bushes (the track frames incorporate the "Fowler" patented articulated suspension), power drives for both front-mounted hydraulic equipment and rear-mounted winch and other units, and substantial mountings as an integral part of the design for fitting bulldozers, enclosed driving cab, etc.

The engine is a 4-stroke direct-injection diesel type, developing 150 b.h.p. at 1,500 revs. per min. To cite a few of its principal features, the capacity is 970cu.in.; "C.A.V." fuel injection equipment is fitted; cooling is by radiator, fan and centrifugal water pump; there is high-pressure forced lubrication; the bore is 5.9in., and the stroke 5.9in. A water-cooled 4-stroke petrol engine is used for starting, its cooling system being integral with that of the main engine, but electric starting is offered as an alternative. The main engine clutch is a dry single-plate type, so constructed as to be free from slip, and readily removable for servicing or inspection without disturbance of other components.

The gearbox, flange mounted to the front of the bevel box, is connected to the main clutch by a short propellor shaft, and contains six forward and four reverse speeds. These are:— forward, from first to sixth, 1.54, 2.3, 3.08, 3.82, 4.45 and 5.8 miles per hour; reverse, from first to fourth, 2.12, 3.18, 4.25 and 5.3 miles per hour. It is interesting to add that this gearbox is exactly the same as that fitted to the makers' 95 b.h.p. "Challenger 3" model, but the unit is in no way overstressed by the increased engine torque, because the tooth loading is the same in both cases owing to the higher input shaft speed of the "Challenger 4" (*viz*, 1,500 revs. per min. against the 825 revs. per min. of the model "3").

The aspect of interchangeability here will not be lost upon fleet-scale operators of such machines, for their requirements in spares may be reduced in

appreciable measure. As to the all-steel fabricated bevel gearbox, this forms a rear cross-member between the frame side plates, and is divided internally into three compartments, the central one containing the bevel wheel and pinion, and each of the outer ones a steering clutch and brake. The bevel box's rear face is drilled to receive such standard auxiliary attachments as a logging winch and power control unit.

The final drives are flange mounted to the sides of the bevel box, and contain a system of epicyclic reduction gears. The units are leak-proof and sealed against dirt by a spring-loaded diaphragm face-seal on the sprocket hubs, while a primary labyrinth-type seal provides a further safeguard. An advantage, already mentioned in respect of the main engine clutch, is that both final drives may be removed for servicing without disturbance, in this instance, of the adjacent track frames. With regard to the tractor's main frame, a box construction is favoured, and comprises robust steel plates running the whole length of the hull, cross braced by the fuel tank, bevel box, and front and bottom sealing plates, and additionally reinforced by the cross beam mountings and top cross stays. The unit, being entirely jig bolted and welded, thus has squareness and rigidity, and, apart from the drawbar mounting, the hull's underside is smooth and free from projections. Ground clearance amounts to 15in.

Turning to the track running gear and suspension arrangements for the "Challenger 4," it should be noted that each track frame is fabricated from two solid steel rectangular sections, and the frames, mounted on a large-diameter rear cross beam, rigidly attached to the hull, are yet free to pivot vertically. The established "Fowler" patent double-cranked axle beam has a notable influence here, the beam being cranked at each end in the form of a "Z" bar and mounted on the hull in bronze bushes, with each end of the beam connected by links to the track frames. In consequence, any upward movement at one side leads to a corresponding downward movement at the other side, oscillation of the beam being controlled by two damper leaf springs, one on each side of the hull. A wide range of movement is therefore possible without detriment to the tractor's stability, while the wear-aggravating tendency of track frames to "toe" in or out is obviated, and all shocks absorbed by the track frames are transmitted to the hull, with no load taken by the final drives.

Six track rollers are bolted to the bottom of each track frame; the roller units are of the centre thrust bushed type, having hardened wearing surfaces, ample grease reservoirs and special dual-purpose seals. Identical and interchangeable with these bottom rollers are the track carrier rollers, two on each side and pedestal mounted. Furthermore, hubs and axles are interchangeable with those of the front idler wheels, the rear driving sprockets, track rollers and idler wheels being the same as those fitted on the "Challenger 3." Single point adjustment for the track tension is another model "4" feature, and totally enclosed spring recoil units are anchored direct to the rear track frame pivot brackets. The tracks used are of the pin and bush type, with all wearing surfaces hardened, and two master pins per track fitted for easier maintenance, while the track plates are of rolled steel section with integral grousers. Alternative widths of track plate are available.

Control of this powerful new diesel crawler should not present any problem to experienced drivers. The main engine clutch is operated by an over-centre hand lever, positioned on the driver's left-hand side, the over-centre operation allowing the clutch to be left in the disengaged position for short periods. Change speed controls, set immediately in front of the driver, comprise a lever moving in a slotted gate for selection of the six speeds, and a separate forward/reverse lever. As a result, where frequent reversals are needed, as in bulldozing, the man in control has merely to move the forward/reverse lever, the speed in reverse being correspondingly higher than in forward, and so time is saved automatically. The engine speed control, it may be added, is operated by dual levers positioned at either end of the driver's seat.

Although the size and scope of the "Challenger 4" will no longer be in question, we should imagine, the principal dimensions may be of interest. In overall measurements, the tractor is 16ft. 1½in. long, 8ft. 4in. wide, and 6ft. 4½in. high (excluding the exhaust pipe and air cleaner). The fuel tank capacity is 75gall., and the tractor's weight in working order is approximately 29,500lb. There is a bottom gear drawbar pull of 28,500lb. in average conditions, and a maximum pull of as high as 33,000lb. in good conditions. The rear internal power take-off shaft is controlled by the master clutch, and the front-end power take-off runs continuously at engine speed. A swinging type drawbar is anchored by a bracket set well forward under the front cross beam housing.

Wider Track-Marshall

A wide-track version of the Track-Marshall 55 has been produced by Marshall Sons and Co Ltd, Gainsborough, Lincs. Known as the 55W, this model has 30-in.-wide track plates which give a total ground area contact of 7,800 sq in. and a ground pressure of 2·4 lb per sq in. The ground contact length of each track is 118 in. There are nine bottom rollers per track.

Powered by a 55-hp Perkins 4-cylinder diesel engine, the 55W is fitted with a hydraulically-operated winch which gives a line pull of up to 23,000 lb. The tractor is 164 in. long and 98 in. wide overall.

For the standard version of the Track-Marshall a new 3-point hydraulic linkage system has been introduced by Marshall's. It has a lifting capacity of 4,000 lb and Category 2 and 3 hitch points. The links are adjustable and the lifting range with the links shortened is from 7 in. to 29 in. above ground level. With the links fully extended the lift is from 14 in. to 36 in. above ground level.

The linkage frame incorporates a swinging drawbar and allows for the operation of pto-driven implements. The provisional basic price of the linkage complete is £428 for the 55 and £458 for the Track-Marshall 70—both prices ex works fitted.

Farm Mechanization November 1967

November, 1951　　　FARM MECHANIZATION

A Twin-cylinder Fowler Tracklayer

A New Marshall product, the Fowler Challenger I, is to be manufactured next year. Primarily intended as an agricultural tracklayer, it is also suitable for land levelling and industrial duties.

Ploughing stubble with the Challenger.

MARSHALL, SONS AND CO., LTD., announce a new tracklayer, the Fowler Challenger I, which will be shown for the first time at the Smithfield Agricultural Machinery Exhibition.

The engine is the Model ED5, manufactured by Marshalls, and is a vertical, twin-cylinder, two-stroke Diesel, water-cooled, loop scavenged, and developing 50 b.hp. at 1,250 r.p.m. Bore size is 5¼ ins., stroke 6 ins. and compression ratio is 16:1. The design aims to combine the simplicity of the ported two-stroke with the quieter running of the four-stroke, four-cylinder system.

High-duty Iron Cylinder Heads

A Monobloc cylinder block in high-duty iron, with separate cylinder heads in similar material, is employed, and a crankshaft with two throws at 180 degrees runs in three main bearings. Each cylinder head contains a pre-combustion chamber, a decompression valve and a fuel injector. A C.A.V. pump of the two-cylinder camshaft type is chain driven from the crankshaft. A mechanical variable-speed governor is attached to the pump, providing steady idling at 350 r.p.m. and governed speed up to 1,375 r.p.m.

The main engine clutch is a 14-in. Borg and Beck single-plate dry type which is mounted directly on the flywheel.

A 5 h.p. Coventry Victor opposed-twin cylinder petrol engine for starting is mounted at the rear of the power unit and above the clutch housing. Starting the Diesel in cold conditions can be eased by running the starter engine for sufficient time to allow the warm circulating water from it to encircle the jacket of the main block.

In working order, the Challenger I weighs approximately 10,900 lb. Overall dimensions are: length, 10 ft. 4½ ins.; width, 6 ft.; and height (excluding exhaust pipe and air cleaner), 5 ft. 9½ ins. The gauge, from centre to centre of the tracks, is 56 ins. and the length of ground contact on each track, 60¼ ins. Width of track plates is 16 ins. and the total area of ground contact, 1,930 sq. ins. Ground pressure is given as 5.65 lb. per sq. in. and ground clearance is 12 ins.

Track frames are coupled at the front by a double cranked axle, oscillations of which are damped by means of two leaf springs.

There are six forward speeds: 1.6, 2.25, 3.11, 3.86, 5.2 and 7.2 m.p.h., and two reverse: 1.27 and 2.96 m.p.h. Drawbar pulls in each gear from first to sixth are quoted by Marshalls as: 10,000 lb., 6,880 lb., 4,500 lb., 3,400 lb., 2,160 lb. and 1,150 lb.

The engine and transmission casings form a robust chassis which is built independently of the track unit. The controls are an integral part of the chassis. The two independent track frames are coupled together at the rear by a large-diameter crossbeam on which they pivot, whilst the forward ends are linked by a double-cranked axle which allows independent oscillation of the frames. Provision for the attachment of loading and earth-moving equipment is made on the frame side channels. Tracks are pin and bush type. Two master track pins each side are provided to simplify servicing.

Each front track idler wheel slides horizontally in guides on the frames and is spanned by the idler fork; between this and the crossbeam pivot bracket is a covered spring which acts as a cushion to the idlers. This arrangement also allows the idlers to recoil should any obstruction be trapped on the inside of the track and passed round the sprocket or idler wheel.

Provision is made for four power take-off drives. A 12-in. belt pulley can be provided for coupling to the rear p.t.-o.

The price of the Challenger I will be announced later.

The ED5 engine

PERFORMANCE CHARACTERISTICS OF THE ED5 ENGINE

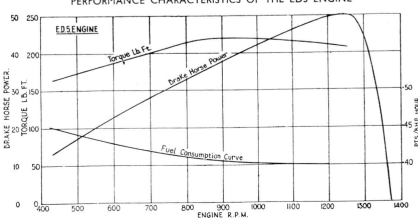

Crawler Tractor Scrapbook Pt 2 — Fowler & Marshall

Maintenance and Repair of the Fowler VFA

After a new lining has been fitted drill it through the remaining holes in the band to dissipate the oil when the brake is applied.

If a brake is short of oil it will be difficult to apply and the steering will be sluggish. If oil has escaped from the casing, one of its escape routes could be via an oil seal between the final drive casing and the main steering or transmission unit. If this is so it would, of course, be confirmed by a corresponding rise in the level of the oil in the main transmission casing. A new oil seal can be installed after the housing (11B) has been removed.

Tracks

The correct adjustment of a track allows the track to be raised not more than 2 in. from the top roller (12A). To adjust slacken the locknut (12C) and turn the adjuster nut (12B) in the required direction, then tighten the locknut.

As the track wears the distance (12D) between the yoke and the guide end will diminish. The minimum distance permissible before the pins and bushes are turned is 1⅝ in. This should be measured when the track is clean and correctly adjusted. Another check against running the tracks too long is the depth of wear on the outside of the bushes where contacted by the sprocket teeth: this should not exceed 0.090 in.

"Fowler" Mark "VFA" Diesel Crawler

Farm Implement & Machinery Review September 1953

"Fowler Mark VFA" (Diesel) Crawler Tractor

Made by John Fowler & Co. (Leeds), Ltd., Leeds, 10.
Date of Test:—May, 1953. Report No. BS/NIAE/53/8.

GENERAL REMARKS

Repairs and adjustments:—None.
Comments:—Engine oil consumption during the 10-hour drawbar test was 4.6pt.
Consumption of coolant:—Negligible.
Use of radiator blind/shutters:—No means of temperature control fitted.
Sea-level (calculated) maximum drawbar h.p.:—Not applicable.

BRIEF SPECIFICATION

Fuel:—Pool diesel oil; specific gravity 0.830 at 60deg. fah.; Cetane No. 56. Nearest U.S. equivalent: commercial diesel fuel.

Tractor:—Serial No. 4704893.

Engine:—"Marshall" series 3A 2-stroke diesel engine, serial No. 4704893; single-cylinder, horizontal; 6.5in. bore by 9.0in. stroke, compression ratio 16:1; "C.A.V." type BKB35S634 injectors fed by "C.A.V." type BPF1B100BS6255 injection pump; gravity fuel feed; two "C.A.V." replaceable element fuel filters; "Marshall" centrifugal governor; governed range of engine speed, 500 revs. per min. to 835 revs. per min.; forced-feed (metered) lubrication from "C.A.V." type BMRDC16L pump. "Marshall" oil-filter with "Vokes" replaceable element; "Marshall" oil-wetted fibre air-cleaner with "Burgess" oil-bath pre-cleaner; thermo-syphon pressurised cooling system with 13¾in. dia., 4-blade fan and twin bank radiator; starting by hand or cartridge with decompression valve and ignition paper; fuel capacity, 12gall.; oil capacity, 7pt.; cooling water capacity, 11gall.

Transmission:—"Fowler" multi-plate, dry, 10¾in. dia. clutch, foot-pedal or hand-lever operated; six forward speeds and two reverse; oil capacity: gearbox 36pt., final drives 38pt. each. Controlled differential steering with hand-lever operated band brakes. Pin-jointed tracks: 1 5/16in. dia. pins, 30 per track; track pitch 6¼in.; 14in. wide trackplates with integral grousers 1⅞in. deep; driving sprocket of pitch dia. 27 3/16in. and 2¼in. face width with 25 teeth; four bottom track rollers per track at 9 9/16in. centres, rolling dia. 8in.; one top carrier roller per track, rolling dia. 8in.; front idler wheel of 21¼in. rolling dia.; distance between driving sprocket and front idler wheel centres: minimum 5ft. 0¼in., maximum 5ft. 3¼in.; track gauge 56in.; approximate length of track in ground contact: 5ft. Suspension by fixed rear crossbeam and cranked front crossbeam. 6-spline, 1⅜in. dia. power take-off shaft; 15in. dia. by 6¼in. face width cast iron belt pulley integral with clutch.

Drawbar:—Swinging drawbar, radius of swing 35¼in.; pivot centre 13¼in. forward of sprocket centre; position of drawbar and power take-off shaft as specified in B.S. 1495/1948, A = 6in., B = 5in. to rear, C = 13in., D = 12in., e = 11 7/16in. offset to nearside.

Nominal speeds (at 750 revs. per min. rated engine speed):—Low ratio: L1 gear 1.26, L2 1.69, L3 2.34 miles per hour. High ratio: H1 gear 2.92, H2 3.90, H3 5.42 miles per hour. Reverse: low 0.96, high 2.24 miles per hour. Belt pulley 750 revs. per min., power take-off shaft 550 revs. per min.

Weight:—Total weight of tractor (with full fuel tank, oil, cooling water and operator: weighbridge figure) was 10,510lb.

Dimensions:—Overall length of tractor, 8ft. 4½in.; overall width, 6ft. 0in.; overall height, 6ft. 10in. (to top of exhaust. (Table p.881).

Crawler Tractor Scrapbook Pt 2 — Fowler & Marshall

In 1957 the Challenger 2 was updated to the Challenger 22 with a Leyland 600 engine. The Challenger 3 became the Challenger 33 in 1962 with a Leyland 680 engine of 125bhp. In the same year the Track Marshall 70 was introduced with a Perkins Six 354 engine giving 72bhp.

1970 saw the amalgamation of Marshalls and Fowlers into Marshall Fowler Ltd. Several new crawler tractor models were announced. Track Marshall 56, Track Marshall 75, Track Marshall 85 and Track Marshall 90.

In 1973 the company announced the closure of the Steam Plough Works at Leeds and the shifting of crawler tractor production to Gainsborough. However this was not the answer to the companies problems and in 1975 the British Leyland Group took over renaming the company Aveling Marshall Ltd.

This new company was short lived and in 1978 the business was sold to Charles Nickerson and be came Track Marshall Ltd.

1986 brought in another move for the crawler tractors, Hubert Flatters took control. Track Marshall of Gainsborough Ltd became the new company. Five Track Marshall crawlers were offered, the TM 110, the TM120, the TM135, the TM140, and the TM155, the figure indicating the hp. Unfortunately Hubert Flatters died in January 1987 and after struggling on for another three years the shareholders sold to the TWR Group in 1990.

At this time, 1990, a prototype Track Marshall 200 was undergoing trials in connection with Waltanna, an Australian company who had patented a crawler tractor with positive drive elastomeric tracks. While the concept of rubber tracks was not new the first few machines ran into problems with the Australian made tracks. The tracks were replaced with Goodyear rubber tracks and the power increased to 210hp but sales did not increase and this model ceased production 1994/95.

In 1997 only the Track Marshall 155 remained on the market.

The Royal National Lifeboat Institution (RNLI) in the UK commissioned several Fowler Challenger IIIs to be built for launching lifeboats into the sea.
There were other makes of crawlers adapted at various times such as Case.
The Challenger III was built in 1953 and several modifications followed. The most important part was the waterproofing of the engine.

Photo: Richard Trevarthen's collection.

Challenger III

Crawler Tractor Scrapbook Pt 2 — Fowler & Marshall

Rotorua, NZ, salesman, Bob Ward, with the hat, delivers a Track Marshall 55 in December 1963. Overloading trucks was not a problem in those days!

Above : This line up of Marshall crawlers belonged to Tom Carson, Nelson, NZ when photographed on the family farm. Left is an early Marshall 55, centre is a Marshall 55 931 series and right is a Marshall 70.

Stephen Arbuckle was showing this Track Marshall 55 at the Canterbury, New Zealand, Easter Rally in 1997. Its serial number, 9312289 makes it a 1967 model. The Track Marshall 55 had a Perkins Four 270D, four cylinder, diesel engine giving 55hp.

FARM MECHANIZATION February, 1956

How the Expert Does It:

Maintenance and Repair of the Fowler VFA

This article deals with the operation, maintenance and repair of the clutch, steering, brakes and tracks of the Fowler VFA tracklayer made by the Marshall Organization. The photographs and information were obtained from the Marshall factory by *Farm Mechanization* **representatives.**

THE VFA is an improved version of the VF which was introduced in 1947 but the improvements do not affect the items discussed in this article, which can, therefore, be applied to the VF also.

The clutch is one of the simplest and most accessible used in a tracklayer. It is combined with the belt pulley and can be dismantled without affecting other components. The steering brakes act on drums connected to a patent epicyclic differential—there are no steering clutches.

Clutch Lubrication

There are two grease nipples connected to the clutch. One (1A) feeds the clutch operating fork shaft and the other supplies the pressure plate ball bearing (1B) on the end of the crankshaft. This latter grease nipple is not shown in Fig. 1 as it is in an end plate which has been removed as part of a dismantling process to be described later. Each nipple should be greased every 50 working hours.

The clutch output gear (6A) and thrust bearing (6B) are lubricated by oil in the transmission casing. Use a reputable brand of S.A.E. 140 gear oil; renew it every 2,000 to 3,000 working hours and keep it up to the level of the plug (1C) in the transmission casing at the rear of the pulley.

The oil inlet is beneath a plate which displays the gear change diagram, and which is in front of the gear levers. The drain plug is in the bottom of the transmission casing to the left of the drawbar centre line as viewed from the rear. Always drain the oil when it is warm and when the tractor is level.

Clutch Pedal Adjustment

If the clutch pedal is kept correctly adjusted *and not used as a footrest*, the clutch should run for a long time without trouble.

The pedal is connected to the clutch operating lever by a cable. When the pedal is at rest, the cable should be free from tension and the rear of the belt pulley should be ⅛ in. from two clutch brake pads which stop the belt pulley when the clutch is disengaged. A corner of one of the brake pads is shown at 1D. The other pad (not shown) acts on the lower side of the pulley.

The ⅛ in. clearance between pad and pulley is adjustable by shims supplied with the tractor. To insert a shim, first slacken the two set-screws which retain the brake pad; the shims are slotted to correspond with these screws and can be pushed in without removing a pad.

Individual components are keyed in the text by a number and letter in brackets: e.g. (1A) means Fig. 1, component A.

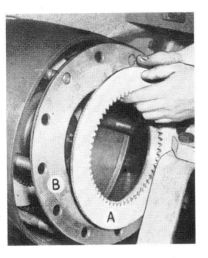

Fig. 3: Removing a clutch lining (A) and a steel disc (B). There are six spare holes in the disc.

To adjust the clutch pedal slacken the rear of the cable nuts (1E) and turn the front nut until about 1-in. movement of the pedal is necessary to tighten the cable when the pedal is depressed. The first firm movement of the pedal should then disengage the clutch and the second movement should bring the pulley into contact with the brake pads.

A lever, interconnected with the clutch pedal, allows the clutch to be controlled by hand from the ground. This lever will also hold the clutch disengaged when pushed fully forward. This is convenient when using the belt pulley but the clutch should not be held disengaged by this method for longer than two or three minutes. At the same time, the engine should be run slowly.

If a clutch slips seriously the belt pulley will become very hot. If this

Fig. 2: Withdrawing the pressure plate (B). The springs should normally be removed before this operation.

Fig. 1: Clutch complete except for a plate which has been removed to reveal bearing and retainer (H).

February, 1956 FARM MECHANIZATION

Fig. 4 (left): Removing the clutch driving centre (B) to examine key (C). The stud (A) is one of six which carry the clutch discs.

Fig. 7 (right): Oil is prevented from leaking into the clutch by a grooved bush (B) and an oil seal (A).

happens, first check the pedal adjustment. If this is correct, the slip may be due to excessive wear of the linings (3A). To renew these, remove the end plate (not shown) so that the bearing (1B) is exposed then remove the six clutch cover nuts (1F), withdraw the cover (1G) and springs (2A).

Remove the bearing retainer plate setscrews and plate (1H) and withdraw the pressure plate (2B) and bearing with a drawer as shown. (Normally the springs (2A) would have been removed before this: they are shown here merely for identification.)

Removing the Clutch Linings

The five clutch linings (3A) and four steel discs (3B) can now be removed. There are 12 holes round the edge of the steel discs. Every other hole engages one of the six studs (4A), the remainder are spare. If the used holes are worn, fit the spare holes over the studs when reassembling the clutch. But before doing this, examine the studs for wear. If this is excessive install new studs after unscrewing the old studs by using two locknuts on the outer end.

Examine the clutch driving centre (4B) for wear at the key (4C). If this is excessive, withdraw the centre and renew the key. When dismantling, note that there is a washer covering the outer end of the key and centre. Finally, keep the linings and discs free from oil or grease.

Major Repairs

The foregoing is all the treatment that the majority of VFA clutches are likely to require but the possibilities of rare troubles cannot be ignored. If gear oil leaks past the crankshaft and down behind the pulley the inner oil seal (6C) may have failed. To renew this is more of an expert's job than renewing the linings. First drain the oil from the transmission casing or else put the tractor so that its front end is about 6 in. higher than the rear: this will drain the oil away from the clutch. Disconnect the clutch control cable and withdraw the operating fork shaft (1I).

Remove the setscrews which hold the clutch back plate (5A) and the clutch brake pads. Then withdraw the pulley and gear assembly, as shown in *Fig. 5*.

Remove an Allen headed screw (6D) from the gear and with a drawer attached to the threaded holes (6E) withdraw the gear and remove the key. The clutch thrust bearing and die can then be removed. (This bearing is shown dismantled in *Fig. 6*.) Next remove the plate (6F) which contains the seal (6C). When fitting a new seal note that the lip of the seal should be towards the gear.

When the plate, gear and yoke are reassembled, it is essential that there is 0.006 in. to 0.008 in. end clearance

(Continued overleaf.)

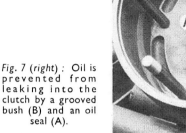

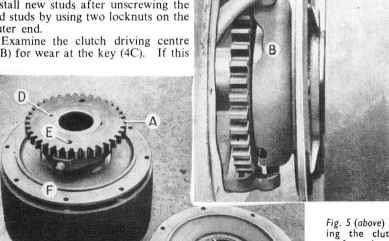

Fig. 5 (above): Withdrawing the clutch complete from the crankshaft.

Fig. 6 (left): After the gear (A) has been removed the plate (F) with oil seal (C) can be lifted off.

Fig. 8 (right): The gaiter (C) on the bottom of the steering lever must be kept in good condition.

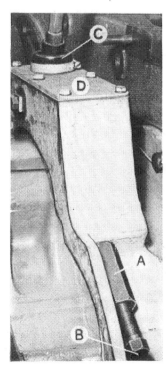

FARM MECHANIZATION February, 1956

Maintenance and Repair of the Fowler VFA
(continued)

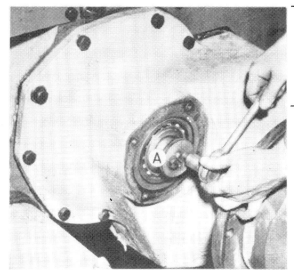

Fig. 9 (left): Removing the bearing sleeve (A) with two setscrews before removing the final drive casing.

Fig. 11 (right): Withdrawing the brake drum assembly (A). The band (C) can then be removed.

between the thrust bearing assembly (5B) and the gear. This clearance is essential because the assembly remains stationary while the gear rotates.

In the extremely rare event of oil having leaked along the crankshaft and into the clutch, it will be necessary to renew the internal oil seal (7A) and the bush (7B) that carries the clutch assembly on the crankshaft. The bush is a force fit in the clutch pulley. It must be reamed or scraped after fitting to give approximately 0.003 in. clearance on the crankshaft. Four lubricating holes should be drilled through existing holes in the pulley. The new seal must be fitted with its lip towards the engine. (Before assuming that this seal has failed, make sure that excess of lubricant in the clutch is not due to an over-enthusiastic use of the grease gun!) Note that the renewal of the above seals and bush should be undertaken by a skilled fitter only.

Steering Brakes

The steering brakes run in oil and should seldom require major attention if treated correctly. When making a turn, apply the brake firmly. To brake in an emergency, declutch and apply both brakes firmly. When a brake is not being used, allow the lever to remain in its fully disengaged position.

The brakes should be adjusted so that each lever has 3½ in.-4 in. free movement at the top. To adjust, slacken the long locknut (8A) and turn the screw (8B) as required, then tighten the locknut. Each brake band is supported by a screw in the housing. In the rare instance of this requiring adjustment, slacken the locknut, tighten the screw, then back it off one turn and tighten nut.

If a brake fails to work properly, first inspect the gaiter (8C) at the bottom of the corresponding lever. If the gaiter is loose or holed, dirt may have entered the brake control chamber. To cure, remove the rectangular plate (8D) clean the chamber then add about a cupful of oil before replacing the plate. (The gaiter should be held with hose clips at the top and bottom —not at the bottom only as shown in *Fig. 8.*)

The brakes are immersed in oil which also lubricates the final drive gears. Use the same grade of oil as for the transmission and change it at the same working intervals. There is a level plug at the rear of each final drive casing.

To get at a faulty brake lining, slacken the corresponding track, remove one of its master pins and draw the track from the sprocket. Drain the oil from the affected final drive casing. Jack the sprocket clear of the track. Remove the sprocket hub bolts and sprocket (another method is to drop the rear bottom roller and leave the sprocket to come away with the final drive casing but it is more convenient to remove the sprocket).

Remove the cover plate from the final drive casing to reveal the shaft end and bearing shown in *Fig. 9.* Remove the two setscrews from the end of this shaft and use two longer setscrews to withdraw the sleeve (9A) as shown. Next remove the setscrews which retain the final drive casing and lift the casing away with block and tackle as shown in *Fig. 10.* Remove the diaphragm plate (10A): this was not included in early models. Withdraw the brake drum and tubular shaft (11A) and then remove the brake band (11C).

Working in Oil

The brake linings are specially made to work in oil and it is important to use the right type. If the wrong linings are used they might swell and permanently brake the tractor.

Fig. 10 (left): Here the final drive casing has been swung away. The next stage is to remove the diaphragm plate (A).

Fig. 12: The tracks are adjusted by moving the front idler wheel forward. This reduces the distance shown (D). The pins and bushes should be turned when this distance has been reduced to 1⅝-in.

Farm Implement and Machinery Review.—February 1, 1959.

N.I.A.E. Tractor Test Report

"Fowler Challenger 22" Diesel Crawler Tractor

MADE by John Fowler & Co. (Leeds), Ltd., Leathley Road, Hunslet, Leeds, 10. Date of Test: July, 1958. Report No. 192/BS; Test No. BS/NIAE/58/8.

STATIC TILTING TEST

The tractor engine was run at rated speed (without load) for a period of 30min. with the tractor tilted to the maximum angle recommended by the manufacturer for continuous operation as follows:—(a) nose up, 30deg.; (b) nose down, 30deg.; (c) offside up, 30deg.; (d) offside down, 30deg. During all these tests no adverse comments were recorded.

GENERAL REMARKS

Repairs and adjustments:—Early in the test the swivel pin in the near-side steering clutch operating mechanism had to be freed because the steering clutch lever did not return to the engaged position; no further trouble was experienced.

Comments:—None.

Consumption of coolant:—Negligible.

Use of radiator blind/shutters:—Not fitted, thermostat control.

Consumption of engine lubricating oil:—For engine test, 0.23pt./hour; for 10-hour drawbar test, 0.25pt./hour.

Soil conditions:—The cohesive strength of the soil as used for the drawbar tests was 19lb./sq. in. U.S.D.A. classification: clay loam.

BRIEF SPECIFICATION

Fuel:—Diesel oil; specific gravity at 60deg. Fah.: 0.836 for engine tests, 0.827 for drawbar tests. Cetane No. 54. Nearest U.S. equivalent: commercial diesel fuel.

Tractor:—(Size in accordance with B.S. 2596) Class II, Group 1, Serial No. 4470130.

Engine:—Leyland Motors, Ltd., type UE.350/81 diesel engine, Serial No. 41; six cylinders, vertical, in-line; 4-stroke cycle, direct injection; 3.96in. bore by 4.75in. stroke, 351cu. in. capacity; compression ratio 16:1; natural aspiration; overhead valves; dry cylinder liners; "C.A.V." type GRVW.B27 mechanical governor, rated engine speed 1,700 revs. per min., governed speed range 450 to 1,850 revs. per min.; "C.A.V." type DFP.6/3 fuel feed pump with one "C.A.V." type BFA.11S2 filter on suction side, one "C.A.V." type F.2/21 deep bowl and one "C.A.V." type F.2/3 paper filter in this order in series, on pressure side; "Leyland" type N.30 injectors fed by "C.A.V." type AAL.6B75/88.EL injection pump, Serial No. R.7493.AZ; forced-feed lubrication from gear-type pump with one "Leyland" full-flow cloth-type element oil-filter, S.A.E.30 lubricating oil; pressurised impeller-assisted cooling system with 19½in. dia. 6-blade belt-driven fan, thermostat for temperature control; electric starter motor, excess fuel device for cold starting, two 12v. "Exide" type 6 KHV 17 JL lead-acid batteries for starting, 24v. 108amp./hour system; fuel capacity, 27gall.; oil capacity, 32pt.; coolant capacity, 5½gall.; "Burgess" type D-8/6562/1 oil-bath air-cleaner with "Burgess" type P.C.-080/1 pre-cleaner outside hood.

Transmission and steering:—"Borg & Beck" single-plate 14in. dia. dry clutch, hand-lever operated; "Fowler" sliding gear type gearbox, six forward speeds and four reverse; clutch and brake steering, hand-levers operating dry multi-plate clutches; foot-pedal operated external-contracting brake bands, ratchet on right brake pedal for parking. Diameters of turning circles: right and left-hand, 15ft. 2in.; rear axle with crown wheel and pinion; spur gear final drive; oil capacities: gearbox, 64pt., final drives, 20pt. each, S.A.E.140 lubricating oil.

Power take-off arrangements:—One rear, 2in. dia. 15-spline giving full engine power, running anti-clockwise viewed from tractor rear, at engine speed.

Sprockets:—Pitch dia. 27 3/16in., 25 teeth, face width 2⅞in.

Tracks:—Non-girder, track pitch 6¾in.; 1 5/16in. dia. pins; 16in. wide track plates, 35 per track; integral grousers 1⅞in. high; track gauge 56in., approximate length of track in ground contact 82in.

Suspension:—Front idler wheels 21¼in. rolling dia.; five bottom track rollers per track at 11¼in. centres, rolling dia. 8in.; one top carrier roller per track, rolling dia. 5in.; suspension by fixed rear cross beam and patent cranked front cross beam, independent of final drive.

Drawbar:—Swinging drawbar, radius of swing 73in., position of pivot centre relative to sprocket centre 48¾in. forward, lateral adjustment 26½in. by five locations, no vertical adjustment; height during test 13¼in. Position of drawbar in accordance with B.S. 2596/1955, A=1⅞in., B=2⅝in., C=3in., D=2in., E=7in., F=13¼in.

Speeds (at 1,700 revs. per min. rated engine speed):—1st gear 1.40, 2nd 2.03, 3rd 2.66, 4th 3.45, 5th 4.50, 6th 6.55 miles per hour. Reverse: 1st gear 1.64, 2nd 2.39, 3rd 3.14, 4th 4.07 miles per hour.

Equipment:—No ancillary equipment fitted.

Weight:—Total weight of tractor (including fuel, oil, coolant and operator: weighbridge figure) was 14,958lb.; horizontal distance of point of balance to rear sprocket centre 36 9/16in.

Dimensions:—Overall length of tractor, 11ft. 5¼in.; overall width, 6ft.; overall height, 8ft. (to top of exhaust pipe); minimum ground clearance, 10¾in. (to bottom of drawbar pivot pin).

Right: The Fowler Challenger 22—developed from the Challenger 2. In addition to modifications to the chassis and ancillary units, a new type of fuel pump, claimed to increase the torque, has been fitted.

Facts and Figures for Mechanics

14.—Fowler Challenger 33 Tracklayer

The following information applies to tractors from serial number 4481602. The number is on a plate attached to the inlet manifold.

Engine: Leyland, six-cylinder, type AU 680/29 diesel. Compression ratio 15·75 to 1. The prefix letters AU stand for automotive unit; the number 680 is the total capacity in cubic inches (5 in. bore by 5·75 in. stroke); the figure 29 denotes detail differences in design compared with other versions and should always be quoted when ordering spares.

Cylinders: These are formed from dry cylinder liners with a bore diameter of 5·0017-5·0025 in. Maximum permissible wear before renewal, 0·020 in. Reboring not advised. Special tool No. 509601 available for extracting and inserting liners, which are pre-finished and do not require honing after fitting. They can be renewed with engine in place. Projection of liner above top of block should be within $-0·000 +0·002$ in. Shims available.

Pistons: Removable from the top, but not from the bottom, of the cylinder block, with the crankshaft in place. Piston ring gaps: top 0·020-0·024 in.; second, third and scraper rings 0·020-0·027 in. Rings should be renewed when gap exceeds 0·100 in. The top ring is chrome plated, the second and third have tapered sides and the scraper rings have straight sides. No oversize pistons.

Gudgeon pin: Hollow, fully-floating and retained by circlips. It is an interference fit in the piston bosses when cold and an easy push fit in the small-end bush. *Do not force the pin in or out when cold.* Before removing or replacing the pin, heat the piston in boiling water. On engines fitted with connecting rods having oil spray holes in the big-ends, fit pistons to connecting rods with the offset combustion chamber in the piston heads on the same side as the oil spray hole in the connecting rod big-end. Fit offset combustion chamber to camshaft side of cylinder block. Small-end bush bore diameter 1·6255-1·6260 in. Bush to gudgeon pin clearance 0·0005-0·001 in. Connecting rods numbered 1 to 6 from front of engine.

Connecting rod bolts: Must be tightened to a total elongation of 0·006-0·008 in. This should be measured with a micrometer. On no account must the nuts be slackened off to bring the split pin holes into line. If the holes do not line up when the correct elongation has been obtained, file the nut slot. After fitting the pin, back the nut off no more than necessary to nip the pin.

Crankshaft: Supported in seven lead-bronze, steel shell, indium-coated, main bearings. The oil holes in the crankpins are drilled eccentrically to reduce centrifugal loading and also to act as sludge traps to protect the big-end bearings. Main bearing clearance should be between 0·0020-0·0040 in. when new bearings are fitted. Renew bearings when clearance exceeds 0·009 in. Big-end bearing clearance should be between 0·0018-0·0037 in. Bearings should be renewed when clearance exceeds 0·008 in. Both big-end and main bearings are pre-finished and do not require to be reamed after renewing. The crankshaft end float is 0·014 in.

Crankshaft bearing renewal and regrinding: Normally by the time the main bearings are ready for renewal the crankshaft will require to be reground and the engine should be stripped. But should it be necessary to deal with one or more main bearings before a major overhaul this can be done with the crankshaft in place. All the bearing caps must be slackened and the cap from the suspected bearing removed. The upper half of this bearing can then be removed with the aid of Leyland tool No. 245872. Tool No. 245869 will also be required if it is necessary to remove the centre main cap.

It cannot be emphasized too strongly that when an owner regrinds a crankshaft and does not have it re-nitrided, extreme care must be taken that an excessive amount of metal is not removed from the fillets by using a grinding wheel with a corner radius less than 0·15-0·17 in., which is the manufacturer's recommended size. If there are any doubts on this point the crankshaft should be re-nitrided after regrinding irrespective of the amount of metal which has been removed. Main and big-end bearing undersizes are available in five steps of 0·010 in. each. The main bearing caps and engine block nuts are stamped with A, B, C, etc., starting from the front of the engine. They must be refitted to their original positions.

Cylinder head: There are two cylinder heads, each covering three cylinders. Use new gaskets if the old ones are not in good condition. Fit each gasket with

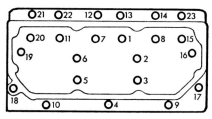

Cylinder head nut-tightening sequence.

the turnover reinforcement surrounding the cylinder uppermost. Do not use jointing compound on the gaskets.

Valves: Stellite-faced with hard chrome-plated stems. The head diameters are 2·20 in. for the inlet and 1·90 in. for the exhaust. The valves are numbered and must not be transposed. When checking valves and springs renew rubber seals if perished. There are two springs per valve. The free lengths are 2·130 in. for the inner and 2·50 in. for the outer. Renew the springs when the inner will compress to 1·25 in. under a load of less than 35 lb, and the outer when it can be compressed to 1·50 in. under a load of less than 74 lb.

Valve stem-to-guide clearances: inlet 0·0025-0·0038 in., exhaust 0·004-0·0053 in. Renew worn parts if clearance exceeds 0·010 in. The valve guides are an interference fit and must be pressed in and out. When correctly installed a guide will protrude 0·520 in. from the top of the head.

Valve seat and face angles are 30° and 29½° respectively. If the seat and valves require to be refaced this should be done with cutting stones of the correct angles. The face of the seat should be concentric with the valve guide bore to within 0·001 in. The face of the valve must be concentric with the valve stem to within 0·001 in. On no account must badly pitted valves and seats be lapped together as this will cause excessively wide seats.

After the valves and seats have been recut they should be lapped lightly together with fine paste to produce a thin seating line towards the top of the seat. Tappet clearances are 0·020 in. for inlet and exhaust (cold).

Valve timing: The inlet opens 10° before TDC and closes 50° after BDC. The exhaust opens 46° before BDC and closes 14° after TDC. Timing marks on flywheel.

Lubrication: Use SAE 20 W in winter (temperature 0° to 30°F), SAE 30 in summer (30° to 90°F) and SAE 40 above 90°F. When the engine is at normal working temperature and speed the oil pressure should be around 60 psi. It must not drop below 5 psi when idling.

Cooling system: The thermostat opens at 165°F and is fully open at 185°F.

Fuel system: CAV type NL 6F 90 injection pump. Firing order 1, 5, 3, 6, 2, 4. Injection begins at 26° before TDC. Timing marks on flywheel.

Electrics: 24-volt system. 108 amp-hr heavy duty batteries. Starting rpm 100. *(The wiring for this machine carries fully annotated labels which obviate the need for a colour code.)*

Engine clutch: Rockford, dry, single plate of 17-in. diameter with Metallo-Ceramic facing and built-in transmission brake. It can be removed without splitting the tractor.

Hydraulics: Lift cylinder diameter 4 in. Cylinder safety blow-off valve pressure 1,500 psi. Pump relief valve blow-off pressure 1,500 psi.

Steering: Free movement of steering levers 3 in. Free movement of steering pedals 2¾ in.

Tracks: Interlocking pin and bush type with three master pins per track. Before removing a master pin, run the tractor on to hard level ground so that one of the master pins is one link above the ground and against the back of the sprocket, then slacken the track and remove the pin. Pins and bushes should be turned when maximum degree of wear on bushes has reached 0·10 in.

Torque wrench settings (in lb ft): Cylinder head: 150-160, mains 215-225, big-end, see *Connecting rod bolts* (column 1) flywheel 112.

September, 1955   FARM MECHANIZATION

New Marshall Tracklayer

The Marshall Organization is to put a new 48 brake h.p., 4-cylinder Diesel tracklayer on the market towards the end of this year. It will be powered by a 4-cylinder Perkins L4 and priced at £1,445.

KNOWN as the Track-Marshall, this new tractor will be made by Marshall Sons and Co., Ltd., Gainsborough, and it will incorporate a number of the features used in the Fowler VF which is made by John Fowler and Co. (Leeds), Ltd., who are also members of the Marshall Organization.

A pre-production model of the Track-Marshall was exhibited at the Royal Show, Nottingham, and a special drawing of this model appears on pages 376 and 377 overleaf. Since this drawing was made, Marshalls have decided to include track carrier rollers. There will be one roller mounted on each track frame to support the underside of the track. Pre-production modifications of this nature are always a possibility, but it is extremely improbable that any are likely to affect the main components and design.

Established Design

One reason for assuming that there will be no major changes between pre-production and production models is that the new tractor represents a combination of engine, transmission and tracks all of which are well established.

Excluding the engine, the Track-Marshall is similar to the Fowler VF but with various components modified, and strengthened where necessary, to apply the extra power—the VF engine develops 40 brake h.p. as against the new engine's 48. The track frames on the VF are of channel steel, those of the Track-Marshall are solid steel of the same overall dimensions. This change has been made to provide stronger anchorage for bulldozers and similar earth-moving equipment.

The double cranked torsion bar which permits the track frames to oscillate vertically while lateral alignment is maintained has been stiffened and installed in stronger brackets. In addition to connecting the track frames, this bar also supports the forward weight of the tractor, as can be seen in the drawing overleaf.

The rear of the track frames are free to pivot on a stationary axle which extends across the chassis immediately in front of the sprockets, thus allowing for the removal of the final drive units for servicing, without disturbing the track frames.

The tracks are the same pin and bush type which have earned a reputation for reliable performance since they were introduced on the VF in 1947.

A foot-operated, 14 in. diameter Borg and Beck single plate clutch transmits the drive from the engine to the gearbox input shaft.

Six Speeds

The gearbox contains six forward and two reverse speeds giving 1.31 m.p.h. in 1st; 1.76 in 2nd; 2.44 in 3rd; 2.78 in 4th; 3.72 in 5th; 5.15 in 6th; 1 m.p.h. in 1st reverse and 2.13 m.p.h. in 2nd reverse.

The gearbox output shaft drives a patent epicyclic differential assembly through which the tractor is steered by two hand-operated brakes.

Each brake acts on a drum fitted to the differential. When a brake is applied, the speed of the drive to the adjacent track is reduced, while the speed of the other is increased. With this type of steering, both tracks are always positively driven during turns. If required for parking, both brakes can be applied simultaneously by a lever which acts through a compensating linkage and which can be locked by a ratchet.

Four-stroke Diesel

The most outstanding difference between the Track-Marshall and the VF is that the engine of the former is a four-cylinder 4-stroke Diesel whereas the VF engine is a single-cylinder horizontal two-stroke Diesel.

The crankcase of the two-stroke engine forms part of the chassis whereas in the new design the front end of the chassis is a U-shaped heavy steel fabrication attached to the main casting which houses the clutch, transmission and differential. The rear of the engine is also attached to the main casting while the front is supported on the chassis by a conventional cross member.

Dimensions

The standard centre-to-centre distance of the tracks is 56 in.—an alternative of 45 in. will also be available. Track shoes, 14 in. wide, are standard and 16 in. and 18 in. wide shoes will be optional. With standard equipment, the ground pressure is 6 lb. per square in.

A rear p.t.-o. is available and provision has been made for the use of a belt pulley, a heavy-duty haulage winch, hydraulic dozer, earth-moving equipment, street plates and electric lighting.

Further details of the specification are: engine, 4-cylinder, bore and stroke 4.25 × 4.75 in.; swept volume 269.5 cu. in. Weight of tractor in working order 10,600 lb. Maximum drawbar pull 12,000 lb. (manufacturer's test). Width 5 ft. 10 in.; height to top of radiator filler cap 5 ft. 1 in.; ground clearance 12 in.

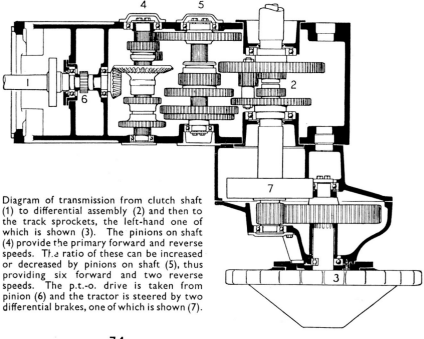

Diagram of transmission from clutch shaft (1) to differential assembly (2) and then to the track sprockets, the left-hand one of which is shown (3). The pinions on shaft (4) provide the primary forward and reverse speeds. The ratio of these can be increased or decreased by pinions on shaft (5), thus providing six forward and two reverse speeds. The p.t.-o. drive is taken from pinion (6) and the tractor is steered by two differential brakes, one of which is shown (7).

Crawler Tractor Scrapbook Pt 2 — Fowler & Marshall

THE TRACK-MARSHALL

KEY TO NUMBERS ON DRAWING

1. Perkins L4 Diesel engine.
2. Clutch housing breather.
3. Parking brake.
4. P.t.-o. lever.
5. Steering levers.
6. Gear lever.
7. High/low range gear lever.
8. Clutch pedal.
9. Left-hand steering brake.
10. Final drive reduction gear.
11. Track recoil spring housing.
12. Double-crank torsion bar.

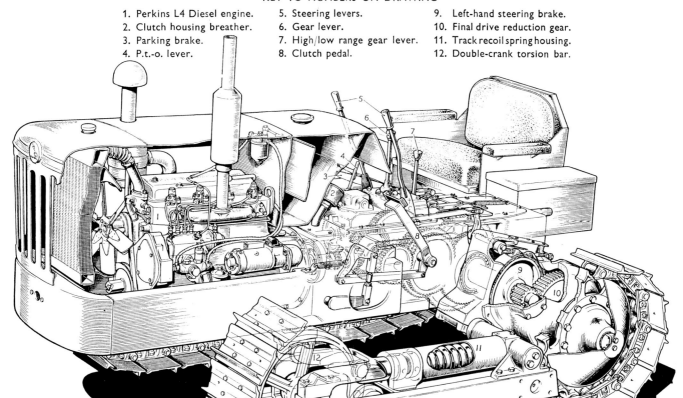

Since this drawing of the pre-production model Track-Marshall was made, the manufacturers have decided to include track carrier rollers. These will be mounted one to each track frame to support the underside of the top of the track. They will be fitted to the production models.

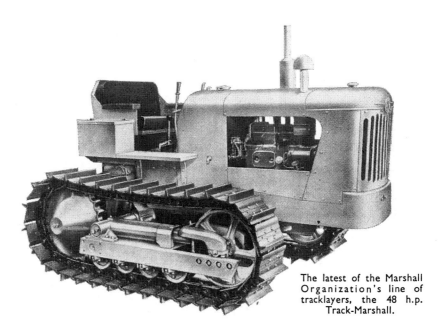

The latest of the Marshall Organization's line of tracklayers, the 48 h.p. Track-Marshall.

Farm Mechanization July 1955

Facts and Figures for Mechanics

12.—Track-Marshall 55 tracklayer

The following information refers to the 55 now in production and is applicable to tractors from serial number 9301651. This number is on the left-hand battery box and also on the corner of the tractor side channel.

Engine: Perkins Four 270D, 4-cylinder diesel. Compression ratio 16 : 1. Maximum governed no-load speed 1,910 rpm.

Cylinders: Wet cylinder liners are fitted, sealed at top by cylinder head gasket and at bottom by rubber rings. Bore diameter 4·250 in. Reboring not permissible. Liners can be removed with engine installed. Honing not required after fitting. When fitted the top of a cylinder liner should not be more than 0·001 in. below top face of cylinder block or more than 0·003 in. above it.

Pistons: Pistons and connecting rods can be removed from top of cylinders with engine installed. Piston ring gaps: compression ring 0·012-0·019 in., scraper 0·012-0·017 in. Top ring chromium-plated and all three compression rings taper faced. There is one oil scraper above the gudgeon pin and another below. The letter F is stamped on the top of each piston and must be fitted towards the front of the engine. Pistons numbered one to four counting from front of engine. No oversize pistons.

Gudgeon pin: This is a fully floating pin which should be fitted after the piston has been heated to about 100°F in water or oil. Small-end bearing bore diameter 1·4382 in. This bearing is a press fit and must be reamed after fitting. Small-end bearing to gudgeon pin clearance 0·0007 in. Connecting rods and big-end bearing caps numbered 1-1, 2-2 etc., counting from front of engine. The numbered sides must be fitted towards camshaft side of cylinder block.

Crankshaft: Three main bearings. Shaft to main bearing clearance 0·0025 in. Crankshaft end float 0·007 in. Reaming not required after mains have been renewed. Crankpin to big-end clearance 0·0025 in. The following main and big-end bearing undersizes are available: 0·010 in., 0·020 in., and 0·030 in.

Cylinder head and valves: Inlet and exhaust valve seat angle 44°. Inlet and exhaust valve face angle 45°. Tappet clearance for both inlet and exhaust 0·010 in. hot. Clearance between valve stem and guide: inlet or exhaust 0·0015-0·004 in. When fitting valve guide press it down to shoulder on guide. When fitting new valves take care that clearance between valve head and cylinder head bottom face is not less than 0·057 in. inlet, 0·053 in. exhaust. Maximum clearance should not exceed 0·140 in., inlet or exhaust. Check this by putting a straight edge across bottom face of cylinder head and measuring distance between straight edge and valve head. Valve spring free length 2⅛ in.

Valve timing: To check, set valve clearance of number four cylinder to 0·025 in. with number four piston at TDC compression. Remove atomizers. Turn engine until inlet push rod of number four cylinder just begins to tighten. This is the point at which the inlet valve begins to open. Check through inspection hole in flywheel housing that flywheel TDC mark is central within inspection hole. If TDC mark is within plus or minus 3° of TDC position the valve timing is correct. Reset valve clearance to 0·010 in. warm.

Lubrication: Use oil with an SAE rating of 20/20W in winter and 30 in summer. Oil pressure 25-50 lb per sq in. with engine at normal working temperature. Oil pump incorporates a detachable relief valve which is screwed into outlet side of pump body.

Fuel system: The fuel pump is a CAV distributor type with mechanical governor. Firing order 1, 3, 4, 2. Injection is timed to 16 flywheel degrees before TDC. Timing marks on pump carrier flange, situated between fuel pump on front timing case, and on pump body flange. There is also a scribed line on the fuel pump driving gear and this should coincide with a line on the triangular driving plate secured to the driving gear hub. Injection pressure 170 atmospheres.

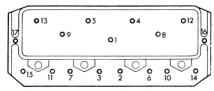

Cylinder head tightening sequence.

Electrics: Two 12v Exide 3XCS21L or Crompton Parkinson 3HC21L batteries. Lucas M45G(SID) 26148B starter motor. Lucas C45P6-22526 dynamo.

Clutch: Borg and Beck dry, single 14-in.-diameter plate.

Hydraulics: Lift cylinder diameter 3·997-4·003 in. Lift piston diameter 3·994-3·996 in. Lift cylinder blow-off pressure 1,000 lb per sq in. Pump flow 13 gal per minute at 1,800 rpm.

Threads: Unified series. Torque wrench settings in lb ft for the following nuts or studs: cylinder head 110 (cold), big-ends 100-105, mains 125-130, flywheel 75.

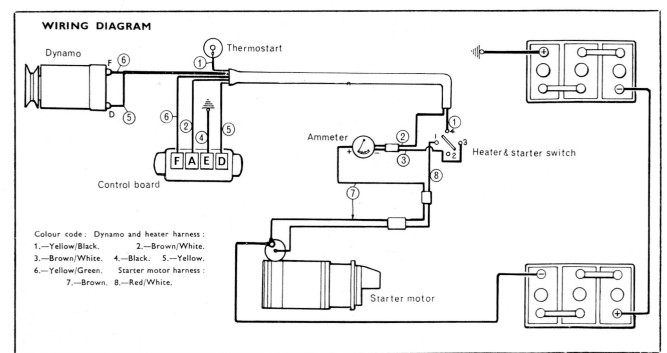

Crawler Tractor Scrapbook Pt 2 — Fowler & Marshall

In 1970 the Track Marshall 55 was updated to the Track Marshall 56 with 57bhp from a 4 cylinder Perkins engine. Stephen Arbuckle had serial number 1183983 on show at the Canterbury Easter rally in 1997.

Below:
A Track Marshall 55 loaded on an International truck and ready for delivery at Rotorua, NZ in 1959.

Bob Ward collection

Track Marshall experts can guess on which model this is.

Actually it is a Track Marshall 70 with TM 75 tinware on the front. Tom Carson, Nelson, NZ is a local farmer who has owned TMs for many years. The New Zealand importing agents for Track Marshall were in Nelson consequently there were plenty of TMs on the local properties which tended to be hilly. Tom has cultivated all the slopes seen in the back of this photograph with TMs.

Crawler Tractor Scrapbook Pt 2 — Fowler & Marshall

Another Nelson, NZ, farmer and contractor who is very keen on Track Marshalls is Jim Stringer. He has owned, TM 55s, TM 56s and now a TM 100. A neighbour has an Aveling Marshall 100.

The Track Marshall 100 was available with a Perkins 6.354 diesel engine. producing 100hp. It had a five speed forward and reverse gearbox. The same machine became the TM105 with a Ford 2714E diesel engine.

Below:
The Track Marshall 200 was made by the Australian company Waltanna.
It was powered by a 200hp, six cylinder Cummins 8.3 litre, turbo-charged diesel engine.
Unfortunately there was trouble with the rubber tracks. In spite of some modifications such as increased power to 210hp and Goodyear rubber tracks sales did not increase. Production was ceased in 1993.
Right: The improved Track Marshall 200

Photo from Richard Trevarthen's collection.

Crawler Tractor Scrapbook Pt 2 — Fowler & Marshall

Getting the Best From...

THE TRACK-MARSHALL

This article deals with the Track-Marshall 55, 70C and 70H tracklayers made by Marshall Sons and Co Ltd, Britannia Works, Gainsborough. Manufacture is at the company's Leeds works where tracklayers were made before the first World War and where they have been in continuous production since 1936.

During a visit to the factory the author concentrated on obtaining information on those aspects of tracklayer operation most likely to suffer abuse and neglect. This information is given here in logical sequence beginning with the adjustment of the engine clutch pedal (an adjustment frequently neglected). The article must not be regarded as a substitute for the instruction book but rather as a reminder of its value. The photographs were taken on the factory assembly line especially for this article, consequently some of the machines are shown stripped. The appearance of these should not be taken to mean that considerable dismantling is required to make the adjustments described.

By T. Hammond Cradock

Fig 1: The tracklayer is at its best in heavy going.

The clutch pedal of the Track-Marshall 55 must have a free movement of 1½ in. As the clutch wears this movement is gradually reduced. If this is ignored the free movement will eventually disappear and the clutch disengaging mechanism will be put under constant load instead of coming into action only when the clutch is disengaged. The result will be unnecessary wear on the disengaging mechanism. This will allow the clutch to slip and destroy itself.

To adjust the free movement of the pedal, slacken the locknut (2A) and turn the adjuster (2B) until the correct setting has been obtained. Tighten the locknut. Some drivers keep a foot continuously on the clutch pedal. This is wrong because it partially loads the disengaging mechanism and causes the same troubles as result from no free movement of the pedal.

The clutch of the 70C and 70H is hand-controlled by a lever as standard. Pedal control is an optional alternative. The lever should have a free movement of 2½ to 3 in. The reason for this is the same as for the free movement of the pedal. To adjust the lever, slacken the locknut (3A) and turn the adjusting link (3B). Then tighten the locknut.

If the clutch lever is used as a handrest make sure that it remains back against its stop or the disengaging mechanism may become partially engaged and wear unnecessarily.

Fig 3: This shows the clutch lever adjuster (B) and locknut (A) of the 70C and 70H.

Fig 2: Track-Marshall 55 clutch pedal adjuster (B) and locknut (A).

Fig 4: Steering levers (A), locknuts (B) and adjusters (C).

Crawler Tractor Scrapbook Pt 2 — Fowler & Marshall

Fig 9: Track removed to show front idler (A), yoke brackets (B) and guide bars (C).

Fig 10: Only an experienced eye can judge the treatment needed to deal with wear at points A, B, C and D.

THE TRACK-MARSHALL
(continued)

the end cover (8B) and then slacken the grease release screw (8D) until this distance is reduced by $\frac{1}{8}$ in. due to the escape of grease. Tighten the screw.

Inspecting pins and bushes

When a track is tightened the front idler wheel (9A) is moved forward. This moves the wheel yoke brackets (9B) on the guide bars (9C) attached to the inner sides of the track frame. When this movement has reduced the exposed part of the guides to about $1\frac{1}{2}$ in. it is time to examine the track bushes for wear.

Wear occurs where the teeth of the sprocket contact the bushes (10A). When the maximum wear is 0·10 in. the pins and bushes should be removed and turned through 180° by a track specialist. This will increase the life of the tracks by about half as much again compared with tracks in which the pins and bushes are not turned. If the wear on the bushes is allowed to exceed 0·10 in. they will not be strong enough to withstand the pressures involved in turning them. Hence the importance of keeping an eye on their condition.

If the wear is allowed to exceed 0·10 in. the most practical thing to do is to leave the tracks in use until the depth of wear reaches not more than 0·125 in. and then to fit new pins and bushes into the old links, provided these have not worn down more than $\frac{1}{8}$ to $\frac{3}{16}$ in. on the rolling path (10B). The original height of all links is $3\frac{3}{8}$ in.

If the link rolling paths are worn more than $\frac{1}{8}$ to $\frac{3}{16}$ in., new chains will be required. But before these are fitted the whole of the track gear should be checked by a Track-Marshall *distributor's* mechanic who will advise whether or not new track rollers (10C) are needed. Expert advice at this stage is most important because it needs an experienced eye to determine the degree of reconditioning required.

When pins and bushes are turned or when new chains are fitted, examine the sprockets. If the driving sides of the teeth (10D) show signs of wear, change the sprockets from one side of the tractor to the other to present new driving faces to the tracks.

Track removal and replacement

Each track has two master pins (11A). These are longer than the standard pins and have a lynch pin through each end. To remove a track, first put the tractor on a hard, level surface with one of the master pins towards the rear of the sprocket as illustrated in Fig 11. Support the back of the track with blocks as shown. Slacken the track. Remove the outer lynch pin (11B) and drive the master pin out with drift and sledge hammer while an assistant holds a heavy object against the inner side of the link containing the master pin.

With an assistant driving the tractor forward slowly, ease the split track off the sprocket with a bar inserted in the bush (11C) from which the master pin has been removed. *During this operation the top of the track can rush forward with devastating speed and force, therefore precautions must be taken to avoid injury.* Do not insert the bar too far into the bush, and keep your body behind the bar. Keep people away from the area until the track has rolled over the front idler on to the ground.

When the track is lying flat on the ground it can be removed either by jacking the tractor clear fore and aft and dragging the track sideways, or by running the tractor forward on to a plank placed in line with the track. If a new chain is being fitted, move the tractor on to this instead of a plank.

Track fitting

Fit a new track in the reverse sequence to that followed for removal. Take note of the following points. The female link (11E) of the chain and the track plate grousers (11D) must be towards the rear of the sprocket as shown in Fig 11. Before replacing the master pin, fit the two collars which are provided for the recesses in the female link. Fit the lynch pins and then adjust track tension.

The tracklayer is designed for heavy loads at relatively low speeds and should not be used with light loads at high speeds. Within reason, the slower the speed the lower the stress and it is better to pull twice the load at half the speed than half the load at twice the speed. Increasing the load also decreases the number of track miles for a given operation.

Alignment of the implement to the tractor is also important and this should be as true as possible so that the outfit tends to run in a straight line. If the implement is badly out of line the tractor will veer to one side. This puts unnecessary strain on one of the steering brakes and should be avoided.

When working in mud do not allow it to pack to the extent of jamming the top carrier roller. In frosty weather clear mud from this roller before leaving the tractor. Park the tractor overnight on firm ground when possible. ∎

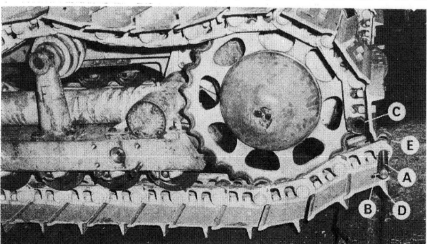

Fig 11: Track ready for connecting. Note the disposition of C, D and E in particular.

Crawler Tractor Scrapbook Pt 2 — Fowler & Marshall

Fig 5: A correctly adjusted track sags slightly at the top.

Fig 6: This track is far too tight. See Fig 5.

This causes excessive wear on the pins and bushes, sprockets and rollers. Compare Figs 5 and 6. Note the slight sag of the track in Fig 5. The tension of this track is correct but the track shown in Fig 6 is too tight. To an experienced eye this is obvious from its taut appearance across the top carrier roller.

Before adjusting a track, first see that it is reasonably clean and that the tractor is standing on hard, level ground. Then, in the case of the 55, slacken the locknut (7A) and turn the adjuster nut (7B) until the required tension has been obtained.

There are two methods of checking tension. The first is to tighten the track until the end cover behind the adjuster nut begins to move back into the recoil spring housing (7C), then slacken the adjuster 10 flats back in a clockwise direction. The second (more practical) method is to tighten the track until about a 2-in. lift of the top of the track can be obtained with a bar (D) inserted underneath as shown in Fig 7.

In the case of the 70C and 70H tractors, the track is tightened hydraulically by pumping grease into the adjuster nipple (8A) as shown in Fig 8. Here again there are two methods of checking track tension. It can be done with the bar as shown in Fig 7 or alternatively as follows. Pump grease in until the end cover just begins to move into the recoil housing at point (8B). Measure the distance between the idler yoke bracket (8C) and

Steering lever adjustment

The 55 and the 70C are steered through a controlled differential governed by two lever-operated brakes. When one of the brakes is applied the differential slows down the drive to one track and speeds up the drive to the other. With this system both tracks are positively driven when the tractor is turning and it is not possible to stop one track no matter how hard the lever is pulled.

The 70H is steered through two clutch-and-brake assemblies controlled by two brake levers. The clutches are kept in engagement under hydraulic pressure which also automatically keeps the clutches correctly adjusted. When a brake lever is pulled back the hydraulic pressure is cut off from the related clutch which is then disengaged. The lever then applies the brake so that the inner track is stopped while the outer track remains in constant drive.

Each of the steering levers (4A) of these three tractors should have a free movement of 3½ in. This movement is measured at the top of the lever. The method of adjustment is the same for each tractor: slacken the locknut (4B) and turn the adjusting screw (4C) until the correct amount of movement has been obtained, then tighten the locknut. There is an adjuster for each lever.

When driving the tractor do not hold the steering levers back so that the free movement is absorbed, otherwise unnecessary wear will be caused.

Track adjustment

Members of the Track-Marshall service department are unanimous in their opinion that the most serious fault in operation is that of running with the tracks too tight.

Fig 7: Checking track tension with a bar (D). This picture also shows the 55 track adjuster (B), locknut (A) and recoil spring housing (C).

Fig 8: The 70C and 70H track is tightened by pumping grease into nipple (A) and slackened by releasing screw (D).

Crawler Tractor Scrapbook Pt 2 — Fowler & Marshall

Track-Marshall 70

Marshall, Sons and Co., Ltd., are now offering a more powerful version of their well-known tracklayer. Called the Track-Marshall 70, it has a 6-cylinder Perkins Six 354 direct-injection engine developing 72 b.h.p. at 1,700 r.p.m., as against the standard model 55 which has a Perkins four-cylinder Four 270D developing 57.9 b.h.p. at 1,800 r.p.m. The 70 is provisionally priced at £2,350, while the 55, which remains in production, costs £1,750.

The new model is similar in construction and configuration to the Track-Marshall 55, but has been re-stressed and given a different gearbox and heavier tracks. A single-range 5-speed gearbox replaces the two-range 6-speed unit, giving forward travel rates of 1.9 to 5.89 m.p.h. and a single reverse of 2.4 m.p.h.

The tracks are longer, having a ground contact length of 6 ft. 2 in. instead of 5 ft. An extra ground roller has been added, and the standard track plate width has been increased from 14 to 16 in.

An N.I.A.E. Test Report (No. 273/BS) has already been published on a pre-production model of the 70. The report quotes a maximum drawbar horse-power of 61.3 at 12,300 lb. pull in first gear, and a maximum sustained pull of 15,400 lb. (Equivalent test figures for the 55 are 47.9 maximum drawbar horse-power and 14,150 lb. maximum sustained pull.)

The overall length, 10 ft. 9¼ in., is 5¼ in. up on the 55 and the weight is increased by 1,540 lb. to 13,160 lb.

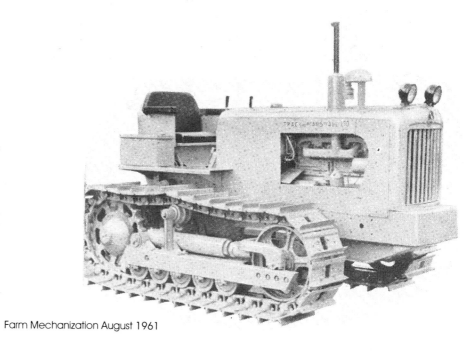

Farm Mechanization August 1961

Jim Stringer, Nelson, NZ puts a Track Marshall 56 through its paces for a local publicity photograph session in 1974.   Photo Jim Stringer Collection.

Crawler Tractor Scrapbook Pt 2 — Fowler Track Marshall

Fowler Marshall & Track Marshall

Model	Production Start - Finish	Comments
Marshall 18-30	1933 - 37	fitted with Roadless tracks, 30hp, 1 cyl.
Gyrotiller	1927 - 39	150hp, 1933 - 170hp, also a smaller 80hp and 30hp model, 6 cyl.
10/70	193 -	70hp, 6 cyl.
10/80	1935 - 38	80hp, 6 cyl.
25	1933 - 34	25hp, 3 cyl.
3/30	1935	25hp, 3 cyl. became 35
3/35	1935	
3/40	193	
4/40	193	
24	1945 -	24hp,
35	1945 -	35hp, 3 cyl. became FD3
FD2	1945 - 48	28 bhp, 4 cyl.
FD3	1945 - 48	35 bhp, 3 cyl.
FD4	194	4 cyl.
		Howard's Rotary Hoes Ltd take over for 1946. Fowler joins Marshall 1947.
VF	1948 - 52	40 bhp, 28dbhp, single cyl.
VFA	1952 - 57	40 bhp, 33.3dbhp, single cyl.
Challenger 3	1950 - 56	109 bhp, 85dbhp, 6 cyl.
Challenger 1	1952 - 55	50 bhp 2 cyl.
Challenger 4	1953 - 74	150 bhp, 135 dbhp, 6 cyl.
Challenger 2	1955 - 56	65 bhp, 52.7dbhp, 6cyl.
Challenger 22	1957 - 74	53.2dbhp, 6cyl.
Challenger 33	1962 - 74	125bhp, 6 cyl.
Track Marshall	1955 -	45.6belthp, 40.1dbhp, 4 cyl.
Track Marshall 55	1959 - 70+	56.9bhp, 52ptohp, 48dbhp, 4 cyl. Leyland Four 270
Track Marshall 70	1962 - 73	72 bhp, 6 cyl. Perkins 6.354
Track Marshall 56	1970 -	57bhp, 4 cyl. Leyland Four 270, Perkins 4.425
Track Marshall 75C/75H	1970 -	75hp, 6 cyl. Perkins 6.354 (AM75?) 75C=balance power steering, 75H=multiplate clutches and brakes steer.
Track Marshall 85	1970 -	94hp, 6 cyl. Perkins 6.354
Track Marshall 90	1970 -	94hp, 6cyl. Ford or Perkins engine Aveling Marshall set up in 1975
Track Marshall 100	197?	100hp, 6 cyl. Perkins 6.354. Becomes Aveling Marshall 100.
Aveling M.105	1976 -	105bhp, Ford 2714E
Track Marshall 110	1986 -	110bhp, 6 cyl. Perkins 6.354 Charles Nickerson takes over 1978
Track Marshall 120	1986 -	120bhp,
Track Marshall 135	1986 -	135bhp, 6 cyl. Perkins T6.354. Hubert Flatters takes over 1986
Track Marshall 140	198?	140bhp, 6 cyl. Leyland engine. An updated Clallenger 33 AM140
Track Marshall 155	1986 -	155bhp, 6 cyl.
Track Marshall 200	1990 -	200 hp, 6 cyl. Cummins 8.3 litre turbo-charged diesel
		1990 Track Marshall of Gainsborough sold to T.W.R. Group. E & OE

Jim Stringer, contractor, shows the holding power of a Track Marshall 55 in the Nelson District, NZ in mid 1960s.

Photo Jim Stringer Collection

Crawler Tractor Scrapbook
Farm Mechanization Advisory Service

TRACK TENSION ADJUSTMENT
Please let me have instructions for the dismantling of the track tensioner adjustment on the Fowler FD3 tracklayer.

Dismantling is a straightforward job provided the parts are not seized by rust. After separating the tracks, the following procedure should be adopted (the recoil units may be dismantled either in position or taken off the track frames complete):

First, slacken the locking device (part No. Y1537): this is a long rod screwing into a cone which, when tightened, prevents rotation of the adjuster. Next, unscrew the adjuster: this is attached to the end of a tube (Y1531) which at its opposite end screws into a tube nut (Y1543). The adjuster should be unscrewed until it comes right out of the tube nut.

Having fully unscrewed the adjuster, the idler yoke with the container or idler carrier (Y2056) can be separated from the spring bracket (Y1524), thus exposing the internal parts.

October, 1953

BUYING A USED TRACKLAYER
What are the points to look for when buying a second-hand tracklayer? The machine we have in mind is a Fowler VF.

This tractor has a two-stroke engine and it is essential that the crankcase compression be good. If it is poor, the tractor will be difficult to start and its performance will be sluggish. If the engine starts easily when hot or cold, it is fairly safe to assume that the crankcase and combustion chamber compressions are both good. This being the case, the rest of the mechanism is likely to be in similar condition.

Before buying the tractor, we recommend that you use it for at least two hours under full load. If, during this time, it does not overheat, discharge excessive quantities of black smoke, or show signs of seizing up, its condition can be regarded as satisfactory. The manufacturers, John Fowler (Leeds), Ltd., Leeds, 10, will no doubt be able to tell you the age of the tractor, provided you give them the serial number. The safest way of ensuring you buy a worthwhile machine is, of course, to have it inspected by someone who is fully conversant with it before purchase.

December 1956

SECOND-HAND TRACKLAYER
I recently bought, second-hand, an Allis-Chalmers WM tracklayer, with three forward speeds and one reverse. It slips out of low and reverse gears. What is the most probable cause of the trouble? Could the engine of this tractor be supercharged?

The most probable cause of the gear pinions slipping out of engagement is excessive wear on the pinion teeth and shaft bearings. It is not a practical proposition to try to supercharge the engine of this tractor.

QUERY
The engine oil escapes from my International TD9 engine into the clutch housing via the crankshaft rear oil seal.—The crankshaft and rear main bearing are in good condition. I have renewed the oil seal twice, but the trouble remains.—Please can you suggest a cure.

ANSWER
The most probable causes are: (1) displaced oil seal; (2) excessive piston blow-by, due to worn or stuck piston ring and/or worn liners; (3) clogged crankcase breather. In our opinion, causes 1 and 3 are the most common.

The seal, as you know, comprises two strips of felt pressed into housings encircling the crankshaft. Sometimes the drag of the crankshaft "spins" the felts in their grooves and causes joint gaps through which the oil escapes. To prevent this form of displacement, fit the felts as follows: soak them in oil and press them well into their respective housings. Drill an $\frac{1}{8}$-in. hole through each side of each felt retainer groove and felt. Next, rivet the felts in position with aluminium or other soft rivets. (The rivets must be soft otherwise the felt housing may be damaged during the riveting process.)

Cause 2—excessive piston blow-by—must be removed either by new rings or new liners, piston and rings.

Cause 3 can be removed by dismantling the crankcase breather and washing the steel-wool element in clean fuel.

June 1949

TRACK RECONDITIONING
I should welcome your opinion regarding the most economical way of reconditioning FD3 tracks that incorporate 26-in.-wide cast-steel track shoes. The following methods have been suggested to me:—
1. Fit new pins into the oval pin-holes: and obtain approximately three months' running.
2. Cut old track plates so that 1-in. diameter holes can be welded to the original plates and then fit new pins.
3. Bore the pin-holes to take $1\frac{1}{8}$-in. diameter pins.
4. Bore pin-holes, fit $\frac{1}{8}$-in. steel bushes and new pins.

We regard suggestion No. 1 as the most practical method of dealing with these tracks, but point out that when a track has worn two sets of new pins it is generally necessary to renew the track. The steel castings are 13% manganese and are therefore extremely difficult to machine. In our opinion, it would prove uneconomical to try to fit bushes or to enlarge the existing holes; in fact, this has been attempted and abandoned because of the high cost.

October 1954

FITTING TRACK GROUSERS
In the article "What's It Worth?" in your December issue, it is stated that grouser attachments can be fitted to worn track shoes on a TD-6 tracklayer. Could you enlarge on this point and, also, explain how the track shoes could be rebuilt.

The grousers referred to are specially supplied for bolting on to the track shoes: they are approximately 3 in. high and, as a general rule, eight are sufficient per track. Worn grousers can be rebuilt by welding suitable strips of metal on to the worn areas. Electric arc welding would be cheaper than gas welding for this job.

February, 1957

REMOVING TRACKLAYER CLUTCH
Although I have removed all the obvious components from one of the steering clutches of my International TD6, I cannot withdraw the clutch. Can you suggest why?

Maybe you have omitted to compress the main spring. This is done by three $\frac{7}{16}$-in. dia. U.N.C. $2\frac{1}{8}$-in. long studs. They should be screwed into holes provided in the clutch hub plate. When they have been tightened you should have no difficulty in extracting the clutch.

March 1959

FOWLER TIMING CHAIN
Please tell me the correct procedure for fitting a new timing chain to a Fowler FD2 engine, Type DU. The engine is stripped down but the old chain is still in position.

So far as we know no specific instructions were issued by the manufacturer regarding the timing of this engine, so we suggest that you apply the usual method in such circumstances and mark all the components before removing the chain. If these marks are registered relative to each other in their original positions the timing will be correct when the new chain is fitted.

June, 1961

QUERY
I am unable to prevent water leaking from my Fowler F.D.3 water pump. Recently I renewed all the components, but leakage recurred within two days.

ANSWER
As you have renewed all the components, the trouble is probably due to incorrect installation or maladjustment. Did you ensure that all the old packing had been removed before installing a new packing? If any remained, it would probably prevent proper seating of the new packing. Furthermore, the packing ring joints must be staggered round the shaft, otherwise a continuous gap will be formed and allow leakage.

If the packings are correctly installed, the most probable cause of early failure would be overheating due to excessive tightening of the packing gland nut. Evidence of over-tightening will be given if the packing housing becomes unduly warm when the engine is running. To correct this, slacken the nut as far as possible consistent with preventing leakage.

Finally, be sure to use the correct grade of grease: Shell F2 is recommended by the engine manufacturers, and the grease cup should be given a half turn every four hours.

Hofherr-Schrantz-Clayton-Shuttleworth

Clayton and Shuttleworth, Lincoln, England was a manufacturing company that produced steam engines and farm equipment including crawler tractors. They wanted to move into Europe so formed a company in Hungary. In 1912 they withdrew from that market and Hofherr and Schrantz took over becoming Hofherr-Schrantz-Clayton-Shuttleworth or HSCS. While initially the firm made farm machinery, in 1924 the first HSCS tractor appeared on the market. It was powered by a single cylinder semi-diesel engine. This was a hot bulb engine that used almost any kind of liquid fuel such as waste engine oil, similar to the Lanz Bulldog.
In later years the firm produced the Dutra tractors.

It has not been possible to find out how many different model crawler tractors were made under the HSCS banner.

This HSCS crawler was on show at the Booleroo Steam & Traction Preservation Society's 1998 rally.
It is a 1937 model L25, engine number 5771. It had a bore and stroke of 7½ x 9 7/16 inches and 800rpm giving 25hp.
Three forward speeds were 1.75, 2.2 and 2.75mph and reverse 1.75mph. Weight was 7,427lbs.

Crawler Tractor Scrapbook   HSCS

HSCS Model L50 Serial Number 6899 one presumes is a 50hp, single cylinder engined crawler. Again little information is available.
Only one L50 worked in Western Australia and this was photographed at Whiteman Park, Perth in 1997.

This is an earlier HSCS L25 crawler than the Booleroo model having only two forward speeds and reverse. It belonged to Ray Copeland, Ashburton, New Zealand, when photographed in 1989.

HSCS		
Model	Production Start - Finish	Comments
Experimental wheel	1923	single cylinder petrol engine
wheel	1924 -	14hp, 1 cyl. single speed transmission, semi diesel engine
L25 crawler	193?	20/25hp, l cyl. two forward speeds
L25	1936?	25hp, 1 cyl. three forward speeds
Le Robuste	193?	40hp, 1 cyl.
L50	1950?	50hp, 1 cyl.
		Lanz took over most of the stock in 1938.

E & OE

INTERNATIONAL HARVESTER

Prior to 1928 IH wheel tractors were fitted with tracks from outside suppliers such as Trackson, Mandt-Freil (France), Moon Track (halftrack) and Hadfield-Penfield. I.H. produced their first commercial tracklayer in 1928. It was the 10-20 and went out of production in 1931 in favour of the T20 which developed 24 drawbar horsepower. A 45 dbhp model, the T-40 was produced in 1932 and a diesel version, the TD40, came in 1933. The T-35, introduced in 1937 was available with petrol, vaporising oil or diesel engines. A distinctive feature of the IH diesel engine was that it started on petrol.

In 1939 a new range superseded the old and included the T6, TD6, T9, TD9, TD14 and TD18. A giant crawler, the TD24, which developed 161dbhp was introduced in 1947.

British production of IH tractors did not start until 1949. The first British IH crawler was the BTD6 produced at Doncaster in 1953. This was superseded by the new BTD6 and BT6 in March 1955. 1957 saw the introduction of the BTD640 and in 1959 the BTD20. The BTD8 followed in 1960. In 1963 the BTD5 came on the market. Production of the BTD6 continued until 1975, the BTD20-E to 1986(?).

Right above: Roger Horrell's McCormick Deering 10-20 TracTracTor made in 1928. With serial # TT688 it is possibly the oldest surviving IH crawler. It was at the Nelson Vintage Engine & Machinery Club's rally in 1996.
The 10-20 had a four cylinder engine, 4¼ inch bore by 5 inch stroke giving 10-20hp. It weighed 7,350lbs. Three forward speeds were 2, 2¾ and 4 mph and reverse was 2½mph. It was recommended as a two plough tractor.

This McCormick Deering 10-20 was photographed at Pleasant Point, New Zealand in 1987 by Jim Richardson.

Crawler Tractor Scrapbook Pt 2 — International Harvester

On the left is a TD5 and the right a TracTracTor 20. The T-20 first appeared in 1931 and used the same engine as the F20. The 4 cylinder, I head engine had a 3¾ inch bore and 5 inch stroke giving an 18.33 dbhp and 25.31 belthp. Weight was 7,010 lbs. Speeds were three forward 1¾, 2¾ and 3¾ mph and reverse.

By comparison the TD5 had a four cylinder engine giving a 23.49dbhp with a total of ten speeds from 1 to 6.54mph. Weight was similar to the T-20 at 7,155lbs. The TD5 was made in 1960.

The TracTracTor 35 was made between 1936 and 1939. It had a diesel I head engine rated at 1,100rpm. Bore was 4½ inches stroke 6½ inches, giving 42.2 dbhp and five forward speeds from 1.75 to 4 mph. Weight was 11,245lbs.

Next in line is the TracTracTor 40, better known as the T-40 Diesel. This crawler was made from 1932 to 1939 and was the first International Harvester crawler with a diesel engine. The IH made, I head, four cylinder engine had a bore of 4¾ inches and stroke of 6½ inches giving a maximun brake hp of 48.26 and maximum dbhp of 43.05. Five forward speeds ranged from 1¾ to 4mph.

Photographs on this page were taken at Bussleton, West Australia, in 1997. All the machines belong to Gary Brookes of Bunbury, WA.

Crawler Tractor Scrapbook Pt 2 — International Harvester

The International TD5 had an International, 4 cylinder, engine with a 3 3/8 inch bore and 4 inch stroke giving a maximum 29.61 dbhp at a rated 2000rpm. The British BTD5 was made from 1963 to 1967 and had the same engine and transmission as the B414, 40hp. Photographed at Busselton, West Australia, in 1997 this TD5 belonged to Gary Brookes.

This 1952 TD6 was photographed at the Manitoba Threshermen's Reunion at the Manitoba Agricultural Museum, Austin, Canada, in 1998. The International, four cylinder engine was rated at 1450rpm. Bore and stroke was 3 7/8 x 5¼ inches. Weight was 8,585lbs. Five forward speeds ranged from 1.5 to 5.4mph. Belthp was 34.38 and dbhp 26.75.

When the T7, TD7 series began has not been possible to work out but the T7C, TD7C was made from 1969 to 1974. These were replaced with the T7E and TD7E in 1974 and the TD7G appeared in 1987 running on to 1991.

The TD7C had 56bhp, the TD7E 65bhp, and the TD7G 70bhp. Pictured is a TD7E.
It had an IH D-239, 4 cylinder engine with a 3.875 inch bore and 5.060 inch stroke. Displacement was 238.8 cub.inches. It had a full power shift transmission with three forward speeds from 2.05 to 5.77mph and three reverse speeds from 2.42 to 6.72mph

A change of colour to road yellow and we have a International TD8 in Eagle Spares yard, Hamilton, New Zealand, in 1998.

In 1960 a new International crawler tractor was introduced to the world market in Britain, the BTD8 refered to as the TD8, powered by a 60hp diesel engine. It remained in production through various series until 1985.

The International TD9 came onto the market in 1939, went through the TD9 91 series, 92 series and B series, finishing in 1974. Pictured just along the road from the author's home in 1998 is a TD9 B series crawler.

The International, six cylinder diesel, engine in the TD9B, had a bore and stroke of 3 11/16 x 4 25/64 inches. Weight was 12,850lbs. It was available with a manual shift or power shift transmission with either four forward speeds and two reverse or five forward and one reverse, manual or High and Low forward and High and Low reverse, power shift.

Back to the Manatoba Agricultural Museum in 1998. This International TD14 is a 1948 model. Its specifications are 4 cylinder diesel engine, bore and stroke 4¾ x 6½ inches, engine speed 1400rpm, 61.56 brake hp, 51.43 dbhp. Six forward speeds from 1½ to 5¾ mph, two reverse, 1½ and 3½ mph.

Crawler Tractor Scrapbook Pt 2 — International Harvester

This International TD15 was in Eagle Spares yard, Hamilton, New Zealand, in 1998.
It was fitted with an International six cylinder diesel engine with a bore and stroke of 4 5/8 x 5½ inches rated at 1650rpm. Drawbar hp was 67.76. It was made between 1958 and 1962, Series 15C finished in 1990. Forward speeds were six from 1.5 to 5.8mph. Weight was 24,555lbs.

Above is Gary Brookes's TD18 photographed at Bussleton, West Australia, in 1997. Its serial number is 2055 which places it about 1941 for a birth date. Maximum dbhp was 72.38 and brakehp 80.32. The six cylinder diesel engine had a bore and stroke of 4¾ x 6½ inches rated at 1200rpm.

Right is a TD20 series C in Brian Clark's yard in Rotorua in 1998. The TD20 series started in 1958 and was still going in 1988 with the Series G. The six cylinder engine in 1958 had 134ehp and by 1976 was up to 210ehp with an eight cylinder engine.

Crawler Tractor Scrapbook Pt 2 — International Harvester

Above: On show at the West Otago Vintage Club, New Zealand, rally in 1995 was this International TD24 owned by Mr W Ward. It is hoped he did not have far to bring it as it weighs over 20 tons. First produced in 1947 this is a 1954 model. The six cylinder engine gave 148dbhp. The Ferguson TEA20 looks like a toy beside it.

Bore and stroke were 5¾ x 7 inches and the engine was rated at 1400rpm. It had eight forward speeds from 1.6 to 8 mph. Maximum dbhp was 154.05.

Above and Right: Cecil Gilchrist's TD25 photographed in 1998. Cecil is another Rotorua local crawler enthusiast. He was told he could have it if he could shift it. Transport required two loads, one for the tractor and one for the blade. It weighed 52,494lbs, nearly 24 tons. It was made in 1960 with a six cylinder, turbo charged engine rated at 1500rpm. Maximum dbhp was 184.68, bore and stroke were 5 3/8 x 6 inches.

Crawler Tractor Scrapbook Pt 2 — International Harvester

July, 1953 — FARM MECHANIZATION

A British-made International Tracklayer

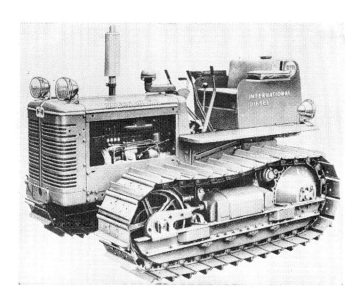

This article describes a new British-made tracklayer which will be introduced at the Royal Show. The tractor, to be manufactured by the International Harvester Co. of Great Britain Ltd., is basically similar to the American-built International TD-6 which was imported into the United Kingdom during the war. The new machine will be available on the home market and for export

A BRITISH version of the International TD-6 tracklayer has been produced by the International Harvester Co. of Great Britain Ltd., in their Doncaster factory. It will be exhibited for the first time at the Royal Show (Stand 112), Blackpool, July 7-10, and delivery is scheduled to begin towards the end of this year. Two models will be available: the BTD-6, with 4-cylinder indirect-injection Diesel engine; and the BT-6, with 4-cylinder vaporizing oil engine.

Prices will be announced by the manufacturers at the Show.

Power Developed

The Diesel model develops 38.2 belt h.p. and 31.3 drawbar h.p., the vaporizing oil model 36.7 belt h.p. and 30.1 d.b.h.p. Further details of each engine are given in the specification on page 280. A drawing of the BTD-6 appears on pages 282 and 283.

Except in the engines, there is no difference between the BTD-6 and the BT-6, and the following are common to each: hand-controlled over-centre clutch; five forward speeds and one reverse; clutch and brake steering; three-point articulated suspension; all-steel tracks with detachable 14-in. wide shoes; 50-in. track gauge centre-to-centre; and ground pressure of 4.62 lb. per sq. in.

Interchangeable Components

Both engines are virtually the same as used in the wheeled Farmall BM and BMD manufactured at Doncaster and major components such as the crankshaft, cylinder head, and cylinder block are interchangeable between the same type of engine whether it be fitted in the wheeled or tracked tractor.

The Diesel engine was first introduced at the 1952 Smithfield Show and it is different from the engine used in the American TD-6 which was imported to Britain during the war. The American engine had a high-low compression combustion system which allowed it to be started and run as a petrol engine until warm. The British-made Diesel incorporates four electrically-heated glow plugs so that the engine can be started from cold as a full Diesel.

Method of Starting

Each glow plug consists of a single coil of wire situated in a pre-combustion chamber adjacent to the injector nozzle. The plugs are connected to the starter battery and controlled by a switch. Thirty seconds is normally the maximum time required for the plugs to reach the temperature necessary for starting. A visible resistance is incorporated in the electric circuit to show when the plugs reach the required temperature; there is also a clockwork device to switch off the current should the driver forget to do so.

(Continued overleaf.)

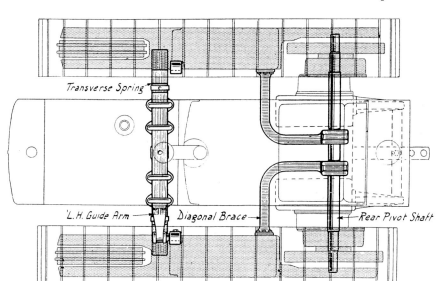

PLAN VIEW OF TRACK SUSPENSION

Each track frame is free to pivot independently on the rear shaft and is kept in alignment by a diagonal brace at the rear and by a roller guide towards the front.

Crawler Tractor Scrapbook Pt 2 — International Harvester

A British-made International Tracklayer (contd.)

The fuel and air system includes an oil-bath air cleaner, two fuel filters, a C.A.V. injector pump with a mechanical feed pump and centrifugal governor. The fuel is gravity fed through a water-trap filter and then to the feed pump which forces it through a replaceable-element filter to the injector pump. A manually operated excess fuel device is provided to facilitate starting and is automatically returned to neutral when the engine reaches 600 r.p.m.

Control of Oil Consumption

The cylinder block contains dry liners that can be renewed without honing or boring. The pistons carry two compression and one oil control ring above the gudgeon pin. A groove is provided below the gudgeon pin so that an extra

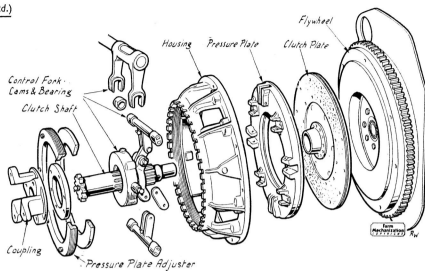

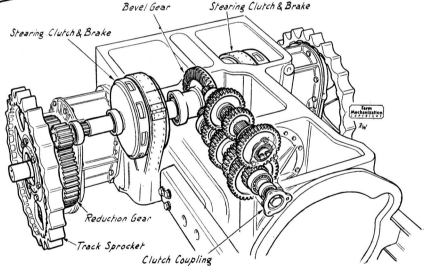

Above: Exploded view of hand-controlled over-centre clutch.

Left: Exploded view of track link and shoe assembly.

Below: Gearbox and final drive.

All drawings in this article are *Farm Mechanization* copyright.

oil control ring can be fitted to limit the oil consumption of a partly-worn engine and thus postpone the need for new cylinder liners.

A three-bearing Tocco hardened crankshaft is used, and the big-end caps are formed so that pistons and connecting rods can be removed from the top of the cylinders. All the crankshaft bearings are pre-finished shells which can be renewed without scraping or boring. They are lubricated under pressure from a gear pump driven by the camshaft.

International tractors have always been respected by mechanics for the accessibility of their major components. The new model maintains this reputation by the fact that the engine clutch, gearbox

(Continued on opposite page.)

GENERAL SPECIFICATION

DIESEL ENGINE
Four-cylinder, overhead valve, water-cooled four-stroke. Cylinder bore, 4 in.; stroke, 5¼ in.; piston swept volume, 264 cu. in.; compression ratio, 16.5 to 1; belt h.p., 38.2; Maximum r.p.m., 1,450.
Starting Equipment : 12-volt electric system, glow plugs, excess fuel device on injector pump.
Cylinder block and crankcase : Monobloc with dry liners; three-bearing Tocco hardened crankshaft; pre-finished, shell type main and big-end bearings; pressure lubrication from gear-type oil pump.
Air, fuel and injection equipment : oil-bath air-cleaner; two fuel filters; CAV injector pump with mechanical feed pump; centrifugal governor and excess fuel device.

VAPORIZING OIL ENGINE
Four-cylinder, overhead valve, water-cooled four-stroke. Cylinder bore, 3⅞ in. stroke, 5¼ in.; piston swept volume, 247 cu. in.; compression ratio, 4.75 to 1; belt h.p., 36.7; maximum r.p.m., 1,450.

Starting equipment : 6-volt electric system; petrol tank.
Cylinder block and crankshaft : as with Diesel engine.
Air and fuel equipment and governor : oil-bath air-cleaner; sediment trap; centrifugal governor.

FEATURES COMMON TO BOTH MODELS
Engine clutch : hand-controlled, over-centre dry, single 12-in. diameter plate.
Speeds in m.p.h. : 1st, 1.5; 2nd, 2.2; 3rd, 3.1; 4th, 3.8; 5th, 5.4; reverse, 1.7.
Drawbar pull in lb. (with Diesel engine): 1st, 8,131; 2nd, 5,514; 3rd, 3,742; 4th, 2,992; 5th, 1,903. (Manufacturer's figures.) D.b. pull with v.o. engine will be similar.
Drive from gearbox to tracks : bevel pinion and gear; multi-plate steering clutches; spur reduction gears.
Steering : lever-controlled multi-plate clutches; pedal-operated contracting band brakes.

Tracks, frames and suspension : all-steel tracks with press fitted pins and bushes and detachable 14-in. wide shoes with integral grousers; channel steel track frames mounted by articulated three-point suspension. Ground pressures, 4.62 lb. per sq. in. (BTD-6) and 4.56 lb. per sq. in. (BT-6).
Weight and dimensions : approx. 7,585 lb.; length overall, 104 in.; width overall, 64 in.; height to top of fuel tank, 59¼ in.; turning radius, 74 in.; minimum ground clearance, 8¼ in.; drawbar height, 12¼ in., drawbar lateral movement at pin, 19¾ in.
Standard equipment : electric lighting and starting; sprocket housing rockshield; exhaust silencer; hour meter; radiator shutter (standard on BT-6 only).
Optional extra equipment : crankcase guard; front pull hook; street plates; belt pulley for rear attachment; front p.t.o. shaft; rear p.t.o. shaft (862 r.p.m.); reduced speed, rear p.t.o. shaft (540 r.p.m.); heavy duty track roller shields; drawbar extension; radiator shutter (BTD-6); bonnet sides.
Manufacturer : International Harvester Company of Great Britain Ltd., Harvester House, 259 City Road, London, E.C.1.

July, 1953 — FARM MECHANIZATION

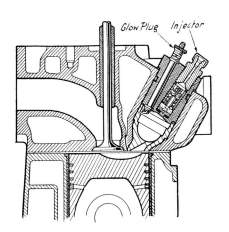

Left: Section through cylinder and combustion chamber.

Right: The Tocco-hardened crankshaft is carried by three pre-finished shell type bearings that can be renewed without scraping or boring.

and steering clutches can be removed and replaced individually without disturbing adjacent components.

The engine clutch is a hand-controlled, over-centre unit with a single, 12 in. diameter dry-plate (see "exploded" drawing). The advantages claimed for this type of clutch are that it can be controlled from any position which allows the driver to reach the lever; there are no pressure springs; the pressure plate is held in its working position by easily adjusted cams, and the complete clutch can be dismantled and reassembled without special equipment.

Transmission and Steering

From the clutch, the drive is transmitted to the gearbox which is part of a single casting forming the main chassis and containing separate compartments for the bevel gear assembly and steering clutches. The gears provide five forward speeds ranging from 1.5 to 5.4 m.p.h., and one reverse giving 1.7 m.p.h.

The bevel gear, which is driven from the gearbox output shaft, is the centre of a built-up assembly including multi-plate steering clutches and two spur pinions that transmit the final drive via two gears that give a reduction ratio of 4.25 to 1.

The steering clutches are controlled by levers, and the steering brakes are pedal-operated contracting bands which act on the outside of the clutch drums. The bands can be removed for relining without disturbing the clutches.

Track Details

The all-steel tracks comprise pins and bushes pressed into links onto which are bolted steel shoes with integral grousers. The pins and bushes can be turned or renewed with the shoes in place.

Each track frame carries the weight of the tractor at the rear by means of a transverse shaft that passes through the main transmission casting and at the front by a laminated transverse spring. Each frame is free to pivot independently on the rear shaft and is kept in alignment by a diagonal brace at the rear and by a roller guide towards the front.

There are four, steel, bottom rollers to each frame. The end rollers have two flanges and the inner rollers have four flanges to guide the track. There is also one track carrier roller and one front idler wheel. Track adjustment is made by moving the front idler along the frame. Two springs are mounted on each frame to allow the idler to recoil against shock or when an object becomes trapped within the circuit of the track.

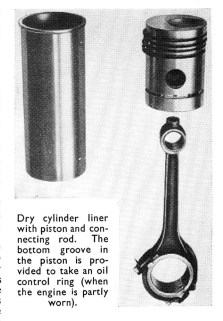

Dry cylinder liner with piston and connecting rod. The bottom groove in the piston is provided to take an oil control ring (when the engine is partly worn).

Below: Accessibility for repairs: the engine clutch can be removed without disturbing the engine or gearbox.

Below: Another instance of accessibility: each steering clutch can be removed without disturbing adjacent components. This photograph shows the right-hand clutch being removed.

Crawler Tractor Scrapbook Pt 2 — International Harvester

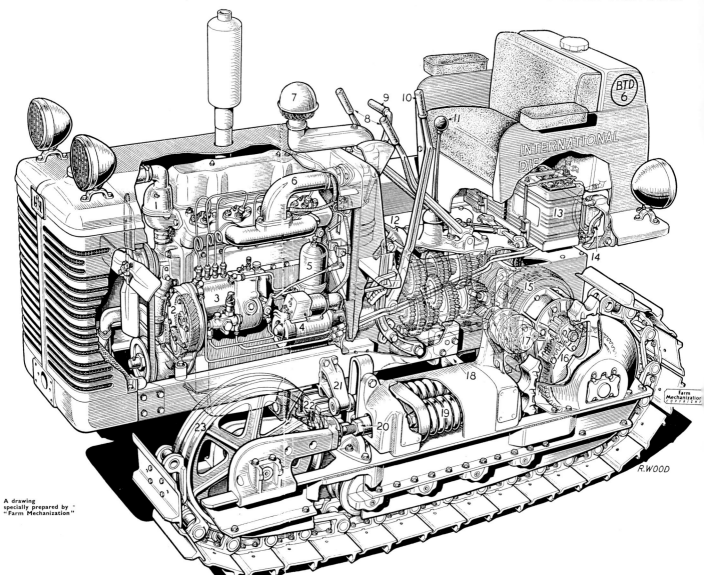

A drawing specially prepared by "Farm Mechanization"

The International BTD-6 Tracklayer

This is a new tracklayer shortly to go into quantity production at the Doncaster factory of the International Harvester Company of Great Britain Ltd. Basically similar to the well-known TD-6 of American manufacture, it will also be available with a vaporizing oil engine (see article and further drawings on pages 279-281).

KEY TO NUMBERS ON DRAWING

1. Thermostat housing
2. Fuel pump drive gear
3. Fuel pump and governor assembly
4. Electric starter motor
5. Secondary fuel filter
6. Air-cleaner connection to inlet manifold
7. Primary air-cleaner leading to oil-bath cleaner
8. Steering clutch levers
9. Throttle lever
10. Engine clutch control lever
11. Gear lever
12. Left-hand steering brake pedal
13. Twelve-volt battery
14. Primary fuel filter and water trap
15. Left-hand steering clutch and brake
16. Final drive reduction gear
17. Track carrier idler roller
18. Left-hand diagonal brace
19. Front idler recoil springs
20. Track adjuster locknut
21. Track frame guide arm
22. Front support transverse spring
23. Front idler wheel

FARM MECHANIZATION — January, 1964

A TRIO from INTERNATIONAL HARVESTER (contd.)

The BTD-5

Features of the BTD-6 and B-414 In New Tracklayer

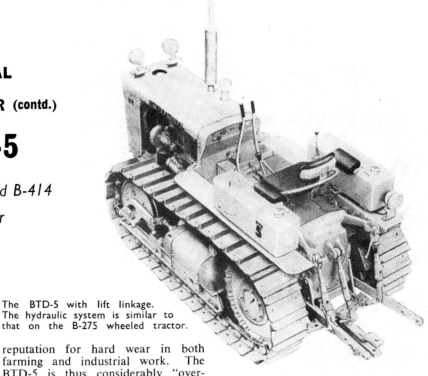

The BTD-5 with lift linkage. The hydraulic system is similar to that on the B-275 wheeled tractor.

IN the face of strong competition from large wheeled tractors and 4-wheel drive models, I.H. has launched the BTD-5 tracklayer aimed specifically at the agricultural market. This action demonstrates the company's faith in the continuing future demand for tracked machines from farmers both at home and overseas, despite anything that rubber tyres can be made to do.

High initial price and the heavy cost of track maintenance are the two main criticisms levelled at tracklayers. I.H. has responded to these by lowering the basic price of the BTD-5 to £1,625—over £200 cheaper than the BTD-6—while retaining the same type of track construction as that of the BTD-6.

These tracks provide more bearing surfaces than the standard pin and bush pattern and have earned a good reputation for hard wear in both farming and industrial work. The BTD-5 is thus considerably "over-engineered" so far as tracks are concerned, the aim being to cut maintenance costs to reasonable proportions.

This machine is the first joint product of the I.H. Bradford and Doncaster works. The engine and transmission up to the final drive are the same as in the Bradford-built B-414. The completion of manufacture is carried out at Doncaster.

Horsepower And Speeds

An output of 40 h.p. is obtained from the BD-154 4-cylinder diesel engine when stripped (36 belt h.p. is quoted for the B-414 and this figure can be taken to apply to the BTD-5). An 11-in.-diameter single-plate clutch is fitted. There are eight forward and two reverse speeds: the first five gears all give speeds of under 3 m.p.h.

"Hands only" planetary steering is operated by a pair of levers, a system introduced for large I.H. tracklayers in the U.S.A. last year. An initial pull on the lever disengages the drive to the track on that side, allowing it to "free-wheel" for a steady turn. Pulling the lever right back applies brake pressure to the track for a tight turn. A pedal gives normal braking and can be latched for parking.

Four basic sizes of track are offered: 40-in. or 48-in. gauge, both with either standard 4-roller or extended 5-roller

Both planetary units operate in one planetary ring gear housing. The planet carrier is splined on to the bull pinion shaft which extends through the sun gear. The steering clutch disc **A** is bolted to the sun gear and the pivot brake disc **B** is splined to the outer end of the bull pinion shaft. An eccentric cam and shaft actuated by the steering lever, operates the brake shoe. (There is a pedal-operated contracting band parking brake between **A** and **B**.) **A** is normally clamped by the brake shoe and prevented from rotating. The first 6-in. of lever travel releases the clamp action, allowing the sun gear to rotate and the planetary system to free-wheel, so that no drive is transmitted and the tractor goes into a gradual turn. Further lever movement, (according to tightness of turn desired) clamps **B**. This prevents the bull pinion shaft from rotating which in turn prevents forward motion of the track so producing a pivot turn.

THE PLANETARY STEERING SYSTEM

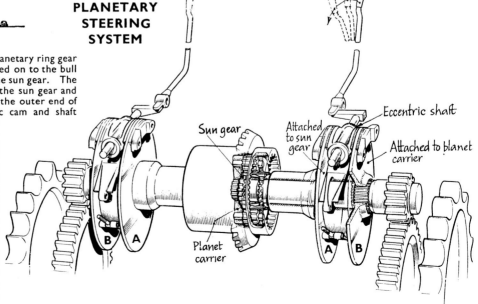

Crawler Tractor Scrapbook Pt 2 — International Harvester

The BTD-5 (contd.)

track frames. Standard shoes are 10-in. wide, but 12-in. shoes are an alternative for both gauges, and 14-in., 16-in., 18-in. or 20-in. for the 48-in.-gauge machine.

A live hydraulic system and Category 2 three-point linkage are available. The hydraulics are similar to those on the B-275 tractor without depth control. A lift of 1,800 lb. is provided by the linkage, for which an offsetting frame can be supplied for use with existing mounted ploughs. *The frame has dual-category hitch pins and can take ploughs of up to three furrows and 900 lb. weight.*

Optional 2-speed p.t.o. gear gives either 540 and 745 r.p.m. or 540 and 1,000 r.p.m., all at full engine power. Speeds are changed by a lever. A belt pulley attachment is driven by the p.t.o.

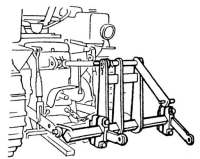

Linkage offsetting frame with dual-category hitch pins.

BTD-5 SPECIFICATION

Horsepower:
Drawbar h.p. (estimated), 31.
Belt h.p. not given (B-414 p.t.o. h.p., 36.7).

Drawbar pulls (estimated) at rated engine speed:
1st gear, 10,600 lb.
2nd gear, 8,800 lb.
3rd gear, 6,300 lb.
4th gear, 4,200 lb.
5th gear, 5,500 lb.
6th gear, 3,300 lb.
7th gear, 2,200 lb.
8th gear, 1,500 lb.

Speeds (at rated governed engine r.p.m. in m.p.h.):
1st (L1), 0.75.
2nd (L2), 1.29.
3rd (L3), 1.83.
4th (H1), 2.20.
5th (L4), 2.74.
6th (H2), 3.75.
7th (H3), 5.35.
8th (H4), 7.99.
Low rev., 1.14.
High rev., 3.33.

Engine: 4-stroke diesel valve-in-head with electric glowplug starting.
IH Model, BD-154.
Number of cylinders, 4.
Bore and stroke, $3\frac{1}{2} \times 4$ in.
Piston displacement 154 cu. in.
Rated governed speed, 2,000 r.p.m.
Maximum torque, 116 lb. ft.
R.p.m. at maximum torque, 1,300.

Electric starting and lighting: 12-volt system, including safety key starter isolation switch, starter motor, glowplugs, resistance indicator, generator, battery.

Clutch: Single plate, dry, heavy duty, 11-in. dia.

Transmission: Selective sliding gear type, carburized alloy steel gears.

Steering: Planetary steering system incorporated in the final drive and actuated each side by a single lever with a two-stage movement.

Final drive (mounted on ball and roller bearings): Speed reduction ratio, bull gears, 4.066:1.

Track dimensions (inches)	4-roller	5-roller
Gauge centre to centre of tracks (basic)	40	40
Gauge centre to centre of tracks (wide gauge attachment)	48	48
Length of track on ground	58.38	70.38
Track shoe width, regular	10	10
Alternative	12	12
Also 14, 16, 18 or 20 in. wide for wide-gauge machine.		
Height of grouser	$1\frac{5}{16}$	$1\frac{5}{16}$
Track pin diameter	$1\frac{1}{4}$	$1\frac{1}{4}$
Track driving sprocket pitch diameter	26.87	26.87
Track shoe bolt diameter	7/16	7/16
Track pin bushing diameter	$1\frac{13}{16}$	$1\frac{13}{16}$
Number of track rollers (each side)	4	5
Number of track idlers (each side)	1	2
Number of track shoes (each side)	33	37

Tractor dimensions:
Length overall (4-roller), 115.5 in.
Length overall (5-roller), 116.6 in.
Width overall (40-in. gauge), 53 in.
Height, grouser tip to highest point, less exhaust pipe, 51 in.
Width over track shoes, 50 in.
Turning radius, 64 in.
Ground clearance under front rigid bar to soft ground line, 10.8 in.
Drawbar height, 10.0 in.
Drawbar lateral movement (either side of centre line), 13.5 in.

Capacities
Cooling system, 18 pt.
Fuel tanks (2), 15 gal.
Engine lubrication, $10\frac{1}{2}$ pt.
Transmission case, 4 gal.
Final drive cases (each), 5 pt.
Air cleaner cup, 1 pt.

Operating weight:
With 10-in. shoes, full fuel, oil, water, without lights, 7,450 lb. (4-roller), 8,030 lb. (5-roller).

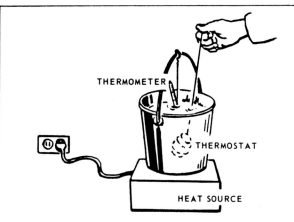

Checking the Thermostat:

1. Suspend the thermostat and a thermometer in a container of water. (Do not allow either one to contact the container sides or bottom.)
2. Heat the water and carefully note temperature when the thermostat starts to open (approx. 80C) and when fully open (88C).
3. If thermostat does not function as described, replace it.

Crawler Tractor Scrapbook Pt 2
International Harvester

The BTD-640—More Details

THE main difference between the new I.H. BTD-640 tracklayer, briefly announced in *Farm Mechanization* last month (page 352) and the BTD-6 upon which it is based is the difference between industrial and agricultural requirements.

In the first place, the stresses and strains inflicted on a tracklayer by industrial use are far more severe, particularly on the tracks, steering and transmission, than those imposed by agricultural use. Secondly, most industrial operations involve the use of heavy, mounted, earth-moving equipment, whereas the agricultural tracklayer works mainly with trailed implements simply attached to the drawbar.

Two Versions—One Model

The industrial tracklayer will do all and more than is required by agriculture but rarely, if ever, can the agricultural model cope satisfactorily with industrial requirements. These differences have led to the general practice of providing agricultural and industrial versions of one model, and this is precisely what I.H. have now done by producing the BTD-640 solely for agriculture and basing it on the BTD-6 which is designed to withstand industrial use.

One effect of the omission of purely industrial features from the new version has been a reduction in price. The BTD-6 is £1,719 and the BTD-640 will be £1,480—a difference of £239 which will no doubt be a very persuasive selling point.

Comparing the specifications of the two tractors, the most important differences are in the engine and track assemblies.

The engine has been derated from 50.5 to 42 brake-h.p. and a lighter track equalizer spring has been fitted. The standard track shoes are 12 in. instead of 14 in. The tracks, frames and rollers and some allied parts have reverted to the less costly design employed on some thousands of the original TD6 type tractor made before the British-built tractor with a 50 h.p. engine was introduced for industrial work.

Further savings have been made by removing the shields which protect the sprockets from rocks, the front wheel weights and other items from the list of standard equipment.

The new version has, of course, a four-cylinder Diesel engine, clutch and brake steering, and steel pin and bush tracks. There are five forward speeds and one reverse, an over-centre main clutch and four bottom rollers to each track. Weighing 8,500 lb., the BTD-640 is 600 lb. lighter than the BTD-6 and it will no doubt prove itself to be a successful medium-sized agricultural tracklayer.

The BTD-640 on standard (12-in.) track shoes.

Farm Mechanization October 1957

New American Crawler Tractor

International "TD-30"

THE International "TD-30" crawler tractor, the largest and most powerful power-unit built by International Harvester Co., is now in production in the United States. This, the latest addition to a long line of construction equipment, is available in both torque-converter and direct-drive models.

The torque-converter model, with 320 engine h.p. at the flywheel, can obtain a drawbar pull of 110,000lb. at 0.5 miles an hour with adequate weight and traction. There are four forward and four reverse speeds, the forward ones ranging from 0 to 3.3 to 0 to 7.3 miles per hour. Reverse speeds are from 0 to 3.3 to 0 to 8.1 miles per hour.

Farm Implement & Machinery Review August 1962

RIGHT

WRONG

Removing the Cable From the Reel or Coil.

Crawler Tractor Scrapbook Pt 2 — International Harvester

August, 1960   FARM MECHANIZATION

Tractor Developments

IT is doubtful if any post-war Royal Show has seen so many tractors with so few changes of design as the 1960 event.

An example of the type of change which designers have made is to be seen in the Ransomes MG-6. This is now green instead of blue, it has fenders over the tracks, a bumper bar is standard equipment, the fuel tank has been repositioned, the tracks, rollers and track frames have been strengthened and the tractor is now known as the MG-40.

New I.H. Tracklayer

Up to the middle of this year I.H. were producing tracklayers at the rate of 3,000 per annum for home and overseas markets. The range comprised a 42-h.p. BTD-640, a 50-h.p. BTD-6 and a 124-h.p. BTD-20. The first model is purely for agriculture while the other two are designed primarily for industrial use but can also be used for heavy-duty agricultural work.

This range has now been extended by the introduction of the BTD-8 which is powered by a 60-h.p. diesel of new design. Priced at £2,150 ex Doncaster

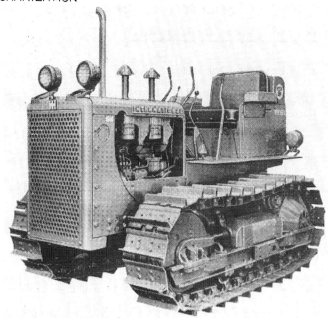

The new International BTD-8 tracklayer is very similar to the BTD-6. Details of new features embodied in this tractor are given in the text on this page.

works, the new tractor is similar in many ways to the BTD-6 but with several new features, in addition to the engine which starts at the turn of a key.

A choice of two transmissions is available. One has five forward speeds ranging from 1.8 to 5.4 m.p.h. and a reverse giving 2.1 m.p.h. The optional transmission has four forward and two reverse speeds and is considered more suitable for shuttle-type operations. The transmission also includes new 30° angle spiral bevel gears, designed to give extra gear mesh area and strength.

A flexible coupling in the drive absorbs shock loads from the track sprockets and thus protects the transmission in this respect. Steering is by the conventional multi-plate clutches with new sintered metal facings.

Improvements have been made to increase the comfort of the driver. A high seat adjustable to four positions is included and the footplate is now level with the fenders. The weight of the BTD-8 is 10,600 lb. The capacity of the four-cylinder engine is 281 cu. in. and the maximum torque is 225 lb./ft. at 1,100 r.p.m.

New "IH" Tractor

Large Industrial Crawler to be made here Next Year

PLANS have now been completed for the production in this country next year of a large new crawler tractor powered by a "Rolls-Royce" diesel engine. It will be made at the Doncaster works of the International Harvester Co. of Great Britain, Ltd., 259, City Road, London, E.C.1, and is intended principally for the industrial and construction equipment field.

It will weigh 13 ton without attachments, and incorporate advanced features of the American "TD-18" tractor seen on the "IH" stand at the Royal Show. It will also involve a considerable expansion at the Doncaster factory, where expenditure on new plant and machine tools will bring the company's investment there to almost £8,000,000. Crawler tractor production capacity at Doncaster, already claimed to be the highest in Britain, will be increased to over 3,000 machines per year. Regular production of the new tractor is to begin next summer.

The engine, a "Rolls-Royce" 6-cylinder diesel type, is rated at 124 b.h.p. at 1,460 revs. per min. The tractor will be suitable for working a 12-16cu. yd. scraper, operating a 3cu. yd. shovel with a break-out force of 27,000lb., and winching loads with a maximum pull of nearly 20 ton at a speed of 77ft. per min. It is intended to make available progressively from British sources the special items of matching equipment necessary, and an early delivery is anticipated of bullgraders, scrapers, cable control units, winches and pipe-booms.

Farm Implement & Machinery Review August 1957

CAUTION: lower blade before making adjustments.

INTERNATIONAL HARVESTER

Model	Production Start - Finish	Comments
10-20 / Model 20	1928-31	10dhp 4 cyl.
T4, TD4	1959-60	T4 21dhp TD4 27dhp 4 cyl.
BTD5	1963-69	40hp. 4 cyl. BD154, diesel.(T4, T5 & diesel Series B, Canada,fits in ?)
T5, TD5,TC5, TDC5	1959-	T5-29.8dhp, TD5-29.61dhp, TDC5 28.5dbhp, 4 cyl.
T6, TD6	1939-56	T6 28.52dhp, TD6 26.75dhp , 4 cyl.
BT6 BTD6	1953-55	31.3dhp 4 cyl.
BTD6	1955-75	50hp 4 cyl., BD 264
BTD6 40	1957-60	31.3dhp 4 cyl. Industrial version of BTD6
TD6	1970-71	4 cyl.
T6, TD6- 61 series	1956-59	42ehp 4 cyl., BD 264
T6, TD6- 62 series	1959-69	TD62A, TD62S, TD62L, 50ehp 6 cyl., C-263, D-282
T7C, TD7C 441 s.	1969-74	56ehp, 4 cyl.
T7E 441 ser.	1974-88	65ehp, 4 cyl.
TD7E 441 ser.	1974-85	65ehp, 4 cyl.
TD8 UK	1960-66	60ehp 4 cyl.
BTD8 80 series	1963	52.1dhp 4 cyl., BD 281, diesel, 10,903lbs
TD8 81 series	1965-	4 cyl.
BTD8 82 series	1965-	58ehp, 4 cyl., BD 281
TD8 USA	1966-75	4 cyl.
TD8- B series	1973-	4 cyl.
TD8 C	1969-82	69ehp 4 cyl.
TD8E	1974-85	78ehp 4 cyl.
T9, TD9	1939-56	T9-46.46bhp, TD9-43.93, TD9 28.97dhp, (1956 TD9 42.6dhp) 4 cyl.
TD9- 91 series	1956-59	56.26dhp 4 cyl., D-350
TD9- 92 series	1959-62	54.5dhp 4 cylinder
TD9 B series	1962-74	75ehp 6 cyl., DT-282
TD12	1981-95	110bhp
T14, TD14	1939-49	T14-60bhp, 1946 only, TD14-61.56dbhp, 54dbhp 4 cyl.
TD14A	1949-55	49.5dhp 4 cyl., D-461
TD14A-141 series	1955-56	76ehp 4 cyl.
TD14A-142 series	1956-58	93ehp 4 cyl.
TD15	1958-62	67.76dhp 6 cyl.
TD15- 150 series	1958-66	105ehp 6 cyl., D-554
TD15- 151 series	1962-	85dbhp, 6 cyl.
TD15- B series	1962-72	120ehp 6 cyl. D-361
TD15- C series	1973-90	140ehp 6 cyl.
TD18	1939-49	53.22dhp 6 cyl.
TD18A	1949-55	67.04dhp 6 cyl., D-691
TD18A- 181 series	1955-56	134ehp 6 cyl.
TD18A- 182 series	1956-58	128ehp 6 cyl.
BTD18	1958.	124ehp 6 cyl., Rolls Royce
T20	1931-39	18.33dhp 4 cyl.
TD20- 200 series	1958-62	134ehp, 111dbhp 6 cyl., 30,300lbs
TD20-201 series	1961-62	113dhp 6 cyl.
TD20B	1963-70	160ehp 6 cyl., DT-429
TD20 C series	1970-75	190ehp 4 cyl.
TD20 E	1977-87	210ehp 8 cyl.
BTD20	1959-74	124 hp 6 cyl.
BTD20- 201 series	1973-78?	135ehp, 110dbhp 6 cyl., Rolls Royce
BTD20- E	19?-86	210ehp 4 cyl.

International Harvester

TD24	1947-54	146.28dhp 6cyl.
TD24- 241 series	1955-59	Gear driven 190-202ehp 6 cyl., D-1091
TD24- 241 series	1955-59	Torque converter 190-202ehp 6 cyl., D-1091
TD25	1960	184.68dbhp, 6 cyl.
TD25 250 series	1959 - 62	230ehp, gear drive & TC, 6 cyl.
TD25B	1962 - 73	230ehp, geardrive, manual & powershift, 6 cyl.,DT-817
TD25C	1973 - 81	285ehp, 6 cyl.
TD25E	1978 - 81	310ehp, 4 cyl.
TD30	1962 - 67	Powershift 320bhp, manual 280bhp, DTI-817
T35, TD35	1936 - 39	T35-35.9dbhp, TD35-42.2dbhp, 4 cyl.
T40	1932 - 39	48.26bhp, 43.05dbhp, 4 cyl.
TD40 TracTracTor	1933 - 39	37.15dbhp, 4 cyl., this was IH's first diesel crawler
T340, TD340	1959 - 65	32.83dbhp, 4 cyl. C-135, D-166
500	1965 - 70	47ehp, 36ptohp, 4 cyl.
500C	1969 - 74	44ehp, 4 cyl.
500E	1976 - 78	44ehp, 4 cyl. E & OE

In 1982 Dresser Inustries Inc. bought the construction machinery division of International Harvester and IH crawlers became part of Dresser's International Hough division. In 1988 Komatsu and Dresser signed a joint venture contract and later Komatsu bought all the Dresser shares. Their crawlers are now marketed by Komatsu America International Corp. It is hoped these machines will be covered in Crawler Tractor Scrapbook Part Four.

A TD9 crushing scrub and light bush in New Zealand in th 1950s.

Crawler Tractor Scrapbook Pt 2

More of Garry Brookes' line up at the Bussleton rally in 1997.
Above an International TD9 pulls a LeTourneau scrapper.

Right an International TD 24.

Below an International TD14A with an Armstrong Holland blade.

Crawler Tractor Scrapbook Pt 2

Experimental tracklayer with rubber tracks on to which the lug plates were riveted. Current I.H. tracklayers have steel pin and bush tracks.

What's new? A rubber tracked TraTracTor beside a Case IH Steiger Quadtrac 1996 vintage. The Quadtrac was photographed at the Manitoba Agricultural Museum, Austin, Canada, in 1998. Unfortunately no date is given with the earlier experimental T20 but the rubber track idea obviously was not proceeded with at that time.
The Quadtrac had a six cylinder powerhouse of 360hp, 12 speed powershift transmission, ranging from 2.2 to 18.7mph forward and 2.9 to 9.1mph reverse. Fuel capacity was 270 gallons. Total weight was 43,750lbs.

Below: A McCormick TracTracTor 20 displayed at the Waikato Vintage Tractor & Machinery Club's Memorabilia Rally, Morrinsville, NZ in 1999.

The LOYD Tractor

I came across this Loyd crawler at the Booleroo Rally in South Australia in 1998. We have a few in New Zealand too.

Captain Vivian G Loyd, managing director of Vivian Loyd and Company, Surrey, England was for many years a partner of Sir John Carden Bt, under the auspices of Vickers Armstrong Ltd. In 1939 he set up his company which throughout the Second World War produced the Loyd Carrier and various prototype military vehicles.

At the end of the war the company concentrated on the production of an agricultural tracklayer, mainly for export. The first production model incorporating many of Carden-Loyd ideas appeared in 1946, nearly 1000 machines being sold by 1950.

The first model, Model P, was powered by a Ford V8 petrol engine, Model D or P/TVO was powered by a Ford V8 engine incorporating a Loyd vapourizing system, for use with vapourizing oil or petrol. Model DP had a Turner 4V95 diesel engine.
In 1950 the Loyd Dragon was launched with a change from differential steering to clutch and brake steering. It was powered either a Turner or Dorman-Richardo engine.

Sales appeared to be booming in 1952 but production ceased during 1953.

Enthusiasts give a Loyd Model DP the once over at the Booleroo Rally South Australia 1998
The Loyd DP is powered by a Turner 4V95 diesel engine developing 30 bhp at 1500 rpm. Drawbar hp is 24.

Crawler Tractor Scrapbook Pt 2 — Loyd

This Loyd Dragon belonged to Jim Richardson when photographed in 1988. The Loyd Dragon was powered by either a Turner or a Dorman engine both developing 36.5bhp. It came on the market in 1950.

Loyd Tractors

Two new Loyd tracklayers were exhibited. The standard Loyd was also shown, with clutch and brake steering as an alternative to the brake-and-differential-type steering unit used in the DP model. Details of the latter appeared on page 106 (June issue of "British Farm Mechanization").

The introduction of clutch and brake steering exemplifies the policy of the Loyd Company to offer many variations of a basic design. These variations include petrol, vaporizing oil and Diesel engines.

The larger of the two new tracklayers shown is powered by a Dorman 4 DWD Diesel engine, quoted as developing 55 b.h.p. at 1,600 r.p.m. It has a variable governor and 24-volt electric starter system.

The engine clutch is a hand-controlled over-centre type. The gearbox gives five forward and five reverse speeds, which are : forward, 1.7, 2.4, 3, 3.5 and 5.1 m.p.h. ; reverse, 1.24, 3.2, 4, 4.68 and 6.8 m.p.h. Steering is by hand-controlled clutches and foot-operated brakes.

The machine weighs 10,500 lb., and its maximum pull is given as 10,000 lb. in first gear. The tracks are the orthodox pressed pin-and-bush type, with detachable steel grouser plates. The rear of each track frame pivots on a transverse axle located in front of the drive sprockets. The front of each frame is anchored to a transverse balance bar, which permits vertical movement of each frame.

General dimensions are : length, 128 ins. ; width, 69 ins. (with 50 ins. track centres) ; and height, 62 ins. The foregoing specification is provisional, and may be modified before full production, which is scheduled for 1950.

The smaller of the new Loyd tracklayers is called the 105 centimetre model, and has been designed primarily for cultivation demanding a narrow-track machine. It has track centres of 31 ins., and is powered by a Fordson four-cylinder vaporizing-oil engine. The machine weighs 4,620 lb., and its maximum drawbar pull is given as 4,500 lb. Carrier-type tracks are fitted, and steering is by differential brakes.

Manufacturers are Vivian Loyd and Co., Ltd., Camberley, Surrey.

Top right: The new Loyd 55 b.h.p. Diesel.
Bottom left: Loyd Model DP.
Bottom right: New Loyd 105 centimetre model.
(Descriptions of these tractors will be found on page 67.)

BRITISH FARM MECHANIZATION   June, 1949

Heavier Loyd Tracklayer

The Loyd Tracklayer combines developments from military and agricultural tracklayer research. It is available with one of three types of engine: Diesel, petrol or vaporizing oil. An outstanding feature is that both track chains and sprockets can be renewed for less than £50.

THE Loyd tracklayer is a product of over 25 years' experience by its designer in the development of track vehicles for military and agricultural use. Evidence of this wide experience is shown by several unorthodox features—a single instance being the track mechanism, which is a direct development of that used on the well-known Loyd Carrier.

To enable an owner to take advantage of the most economical fuel available in his country, the Loyd tracklayer is available as a petrol or vaporizing-oil or Diesel-engined tractor. This has been accomplished by providing a choice of three types of engine.

The P model (petrol) is powered by a Ford V8 petrol engine; the P/TVO (vaporizing-oil) model is fitted with a Ford V8 engine incorporating the Loyd Vaporizing System. The DP (Diesel) model is supplied with a Turner 4V95 Diesel engine.

Starting Arrangements

Electric starting equipment is standard. On the P and P/TVO models this includes a 6-volt 110-amp./hr. battery; while the DP model is fitted with a 12-volt 113-amp./hr. capacity battery. Charging rate is governed by automatic voltage control. A valuable feature of the vaporizing-oil engine is that a starter carburetter incorporated in the main carburetter enables the engine to be started without draining the vaporizing-oil system. The Diesel model is a cold starter.

Engine torque is transmitted by a single-plate clutch to a four-speed and reverse gearbox, whence it is transmitted through a differential to two rear-drive track sprockets. Steering is by two double leading-shoe Girling brakes acting on the differential.

The track chains are fitted with bolted-on grouser plates. An interesting feature of the track chain assembly is that its design permits the removal of links when normal wear has absorbed the range of adjustment provided by a single adjuster operating behind each front idler.

Track Frame Assembly

Bottom rollers are mounted on ball bearings. The top and front idler rollers are rubber shod to resist wear and reduce track clatter. The cost of a set of new chains and sprockets is approximately £42. Each track frame assembly pivots on a transverse axle extending through the chassis immediately in front of the drive sprockets. The front of each track frame is anchored to a transverse balance-bar, which pivots under the radiator. This permits vertical movement while preventing track-frame misalignment.

Special equipment includes: bull/angle-dozer with a 5-ft. wide by 2-ft. high blade adjustable to work on either side at an angle of 26 degrees, winch, belt pulley.

The manufacturers are Vivian Loyd and Co. Ltd., Bridge Road, Camberley, Surrey. Export department, 1, Balfour Place, London, W.1.

The drawbar can be used either fixed or swinging. The power-take-off shaft also acts as belt pulley drive shaft; the pulley clamps onto the shaft by means of a split hub.

SPECIFICATION, "DP" MODEL

Weight: 6,025 lb.
Maximum drawbar pull: 6,000 lb.
Drawbar: Height, 15 ins.; swing from centre, 10 degrees.
Engine: Turner 4V95 four-cylinder Diesel, developing 30 b.h.p. at 1,500 r.p.m. Fitted with variable governor and electric and hand starting.
Drawbar horse-power: 24.
Fuel consumption: 1¼ gallons per per hour average.
Track centres: 43 ins.
Width of track: 14 ins.
Maximum speeds:
 1st—0.75 m.p.h.
 2nd—1.5 m.p.h.
 3rd—2.8 m.p.h.
 Top—4.9 m.p.h.
 Reverse—0.6 m.p.h.
Price: Net f.o.b. London, £895 without p.t.-o.; £950 with p.t.-o.

The DP model fitted with a Turner Diesel engine. The steel plate welded to the front of the track frame is part of the assembly which permits vertical track oscillation whilst ensuring frame alignment.

THE LOYD DP TRACTOR

A drawing specially prepared by
BRITISH FARM MECHANIZATION

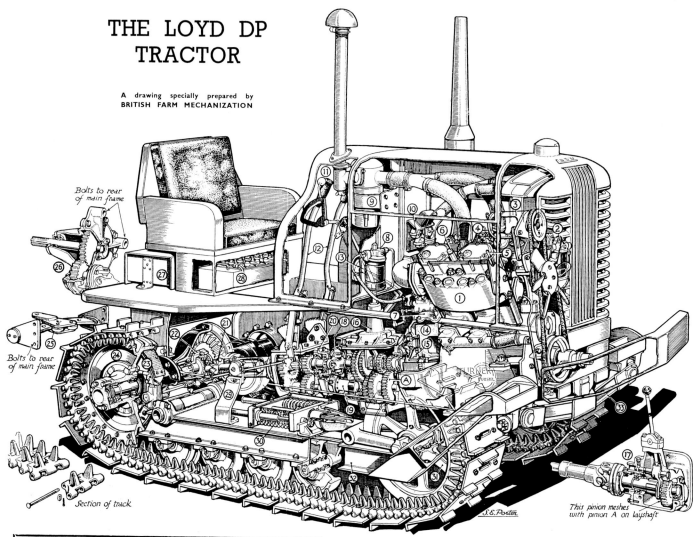

KEY TO NUMBERS ON DIAGRAM

1. Turner, 4-cylinder, Vee-form 33 b.h.p. Diesel engine. 2. Lubricating oil filter. 3. Dynamo. 4. Fuel injectors. 5. Decompressor. 6. Primer. 7. Fuel injector pump 8. Fuel filter. 9. Air cleaner. 10. Ammeter, oil gauge and thermometer. 11. Governor control lever. 12. Steering brake levers. 13. Engine clutch pedal. 14. Flywheel. 15. Clutch with flexible rubber centre. 16. Four-speed and reverse gearbox. 17. Power take-off gearbox. 18. Universal joint. 19. S.U. fuel pump. 20. Secondary gearbox. 21. Crown wheel and pinion. 22. Differential. 23. Girling 2LS brake. 24. Brake drum and sprocket. 25. Swinging drawbar. 26. Power take-off gearbox (rear). 27. Fuel tank. 28. 12-volt battery. 29. Track idler roller support. 30. Track tensioner spring box and adjuster. 31. Track idler wheel. 32. Track frame. 33. Balance beam suspension.

GENERAL SPECIFICATION

Weight (approximately): 6,025 lb. Gears: 4 forward, 1 reverse. Speeds (m.p.h.): .95, 1.9, 3.5, 6.0. Reverse, .76. Track shoe width: 14 ins. Track centre: 43 ins. Length: 124 ins. Width: 61 ins. Maximum drawbar pull: 6,000 lb. Maximum drawbar h.p.: 26.

ENGINE SPECIFICATION

Turner 4V95 Mark 1. Vee form, four cylinder, water-cooled, overhead valve Diesel. Bore: 3.75 ins. Stroke: 4.5 ins. Swept volume: 198 cu. ins. Compression ratio: 16.1. Governed to 1,600 r.p.m. (maximum). Maximum b.h.p.: 33.

Be sure to rest the blade on the ground before leaving the Bulldozer.

TRACTOR SERVICE

March, 1950 — BRITISH FARM MECHANIZATION

No. 10. The Loyd DP

This article deals with the Loyd DP tracklayer powered by a Vee-form 4-cylinder 33 b.h.p. Diesel engine. It is also available as a model D, with a 50 b.h.p. Ford V8 engine for use with vaporizing oil or petrol. The model covered by this article has differential brake steering, 4-speed and reverse gear box and carrier type tracks. Its total weight is 6,025 lb. Extra optional equipment includes bulldozer, rear mounted winch and power take-off.

Cet article traite du tracteur Loyd DP (type à chenille) actionné par un moteur Diesel 4-cylindres en forme de " V " donnant 33 h.p. au frein. Il est aussi livré en modèle D de 50 h.p. au frein, avec moteur Ford V8, pour emploi avec huile vaporisante ou essence. Le modèle dont il est question dans cet article a une direction à frein différentiel, une boîte de vitesses, à 4 vitesses, marche-avant et arrière, et des chenilles du type-transport. Un equipement supplémentaire, au choix, comporte : un " bull-dozer," un treuil monté à l'arrière et une prise de force motrice.

Este artículo trata del tractor tipo oruga Loyd DP impulsado por un motor Diesel en forma de V., con 4-cilindros y de 33 h.p. al freno. También se halla disponible como un modelo D, con motor Ford V8 de 50 h.p. al freno para usarse con aceite vaporizante o gasolina. El modelo cubierto por este artículo tiene gobierno de freno diferencial, caja de cambios de 4 velocidades y marcha atrás, y orugas del tipo portador. Hay equipo extra discrecional que incluye " bulldozer," guinche montado en la parte trasera y toma de fuerza.

THE Loyd DP is a differential-steered tracklayer powered by a 33 b.h.p. Turner Diesel. It is also available, as the model D, with a 50 b.h.p. Ford V8. The Ford engine is supplied for use on petrol or vaporizing oil fuel according to customers' requirements. This article deals only with the Turner Diesel-engined model.

The engine is a Vee-form four-cylinder unit known as the 4V95 Mark 1. It is a water-cooled, overhead-valve, four-stroke unit incorporating dry liners, electric starting, pressure lubrication and centrifugal governor.

The cooling system includes detachable radiator core, water pump and belt-driven six-bladed fan. The water pump is situated at the rear of the engine between the cylinder banks and is shaft-driven by a fan spindle extension. Both pump and fan are lubricated by oil cups, each of which should be filled with engine oil every 40 hours. This is the only lubrication required by these units.

The fan belt is of the detachable link type. Correct tension is obtained by removing a single link when slackness becomes excessive. The water pump impeller shaft carries a self-adjusting carbon gland washer, which requires no attention between major overhauls, when it should be examined and renewed if badly worn.

Air Cleaner

The engine air cleaner is a three-stage, oil-bath type, the correct servicing of which is essential to engine efficiency and long life. In the first stage the air enters a pre-cleaner affixed to the top of the primary inlet pipe, which is vertical. The pre-cleaner is slotted to impart a swirling motion to the air. This motion flings the heaviest particles of dirt away from the primary pipe inlet. The air is then drawn down the primary pipe. At right-angles near the bottom of the inlet pipe is connected a pipe leading to the cleaner body. The air enters this pipe, but, owing to momentum, the heavier particles of dirt cannot negotiate the 90 degree corner. Instead they fall into a pocket formed at the bottom of the primary inlet pipe by a detachable disc.

The air passes down through the centre of the cleaner body until it reaches the oil bath, where it turns through 180 degrees. This change of direction leaves the remaining dirt entrapped in the oil while the air continues upwards through an oil-soaked wire mesh filter into the inlet manifold.

Keep the pre-cleaner slots free from dirt, which would restrict air flow and cause loss of engine power. Clean the dirt trap at the bottom of the primary air pipe approximately once a month. Replenish the oil in the cleaner base every 10 hours in clean conditions and every five in " dust-bowl " conditions. Wash the screens in clean fuel each time the engine is decarbonized. Finally, ensure that cleaner-to-inlet manifold connections are absolutely airtight, otherwise dirty air will by-pass the cleaner and enter the engine.

Fuel System

The commonest causes of Diesel fuel system troubles are dirt, air and water. Dirt damages the finely finished components of the pump and injector nozzles. It can be eliminated largely by always filling the fuel tank through a fine-mesh funnel. Air causes trouble by producing air locks : these can be obviated by not allowing the tank to run dry. Water causes corrosion, and is usually the result of condensation inside the tank. This can be prevented by leaving the tank filled when the tractor is to remain idle.

A gauze filter is included in the tank outlet. If the foregoing precautions are observed, this filter will require attention only during major overhauls. From the tank the fuel is drawn up by an electrically operated S.U. type L pump mounted to the right of the gearbox on a cross-member of the chassis. Access to the pump is by removing the right-hand footplate.

Apart from periodic cleaning of a filter which can be removed by unscrewing the hexagonal brass nut to be found at the bottom of the pump, this unit should require no running attention. The fuel is forced by the pump through a Vokes filter assembly containing a washable filter element, which should be washed in clean fuel every 500 hours and renewed every 2,000 hours.

The fuel continues from the filter to two C.A.V. two-cylinder injector pumps mounted on the rear of the engine behind the cylinder banks. The pumps are automatically lubricated and require no running maintenance. A centrifugal type of governor controls them to settings predetermined by a hand control lever. An adjuster is included in the control lever cable to absorb stretch. The setting will be accurate when the correct high and low engine speeds can be obtained with the minimum of cable backlash. The governor, which limits engine speed to a maximum of 1,600 r.p.m., is entirely automatic and requires no adjustment.

Injector Nozzles

The injector nozzles should be removed, dismantled and cleaned as a first probable cause of misfiring or sluggish engine performance. A set of tools is exported with each tractor, and these should be used in

Left-hand side track drive assembly showing (A) track half-plate included to allow track to be shortened ; (B) grooved rear roller; (C) half-axle retainer flange ; (D) track frame pivot ; (E) track adjuster.

Tractor Service: the Loyd (contd.)

preference to makeshift tools, which might damage the fine finish of the nozzle components. Ensure scrupulous cleanliness when dealing with these components, otherwise damage will be inevitable.

Clean each nozzle as follows (see drawing below) : (1) remove nozzle cap nut with spanner No. E.T.116; (2) see that the valve lifts freely, and then clean surfaces A, B and C; (3) clean oil channel and oilway D; (4) remove carbon from valve seat with brass scraper No. E.T.070 ; (5) clean carbon from gallery E with Tool E.T.071 ; (6) probe spray hole with pricker provided ; (7) brush gently the nozzle valve, seat and cone with wire brush ; and (8) rinse all components in clean fuel and assemble wet.

To prime the nozzles, replace them in the engine and connect, but do not tighten the nozzle supply pipes. Put the pump rack rod in the start position and turn the engine crankshaft until bubble-free fuel exudes from the delivery pipe nipples, which should then be tightened. Correct action of a nozzle will be indicated if it gives a distinct but intermittent buzz when the crankshaft is turned. Do not turn the engine unnecessarily, otherwise the cylinders will become flooded with fuel.

Should the fuel tank be allowed to become empty or if the main supply has been disconnected, it will be necessary to bleed the entire system of air. To do this, fill the fuel tank and remove the nozzle delivery pipes from the pump. There are four non-return delivery valves in the top of the pump. Each comprises holder, spring and valve. Remove one assembly and put the pump rack rod in its start position, switch on the electric pump until a column of air-free fuel flows, replace the valve assembly and repeat the procedure individually with the three remaining valves. Then replace the nozzle delivery pipes and prime them as recommended.

Lubrication

Engine lubrication is by force-feed from a plunger-type pump located between the cylinder banks towards the front of the engine. Correct oil level is shown by the upper of two marks on the dipstick. Use a high-grade viscosity SAE 30 oil, and renew it every 250 hours under normal conditions, or, in very dusty conditions, every 100 hours.

The engine drain plug is screwed into the front of the sump. Draining will be easier if the rear of the tractor is positioned slightly higher than the front. The engine oil filler cap contains a crankcase breather element, which should be rinsed in clean fuel every 1,000 hours.

There are two filters in the engine oil system. The first is located in the sump adjacent to the oil drain plug. Provided clean oil is used, this filter should require cleaning only twice a year. It is retained by an oval flange held to the sump by two hexagonal studs. The second filter is a Vokes full-flow assembly containing a washable element, which should be cleaned thoroughly every 500 hours and renewed after 2,000 hours' use. The correct oil pressure is approximately 15 lb. per sq. in., and is registered by a calibrated gauge mounted on the instrument panel.

Electrical Equipment

The electrical equipment of the Loyd DP comprises 12-volt Lucas C39P dynamo with R.F.97 voltage regulator, Exide 113-amp.-hr. 12-volt battery, and Lucas M45G starter motor with solenoid starter switch.

Rear power take-off and drawbar which can be locked or remain free to swing.

The dynamo is belt-driven from the fan hub. Belt adjustment is made by pivoting the dynamo on its anchor bolts. Correct belt tension allows ½-in. deflection of the belt between the pulley centres. Dynamo lubrication is by a single oiler screwed into the rear end plate. The oiler contains a wick, which should be soaked in oil every 200 hours.

Inspect the battery frequently and maintain the electrolyte ⅜ in. above the plate level by the addition of distilled water. Clean the terminals occasionally and keep them smeared with petroleum jelly. Remove the rubber cap from the starter switch approximately every 500 hours, and smear the plunger with a light coat of petroleum jelly.

Decarbonizing and Tappet Adjustment

The correct clearance of both inlet and exhaust valve tappets is .012 in. when hot. Inspect the tappets every 500 hours and check the gap with the feeler gauge which will be found clipped to a rocker shaft. Be sure that the piston in the cylinder concerned is at top dead centre of its compression stroke before adjusting the tappets, and check the clearances after the adjuster screw locknuts have been tightened.

The engine should be decarbonized and the valves and seats refaced every 1,000 hours. Deal with one head at a time, and always adjust the tappets twice, once before starting up and again when the engine is hot after the head nuts have been finally tightened.

Engine Clutch

The engine clutch is a single dry-plate assembly mounted direct to the flywheel. The throw-out bearing is self-lubricated and requires no further attention. The

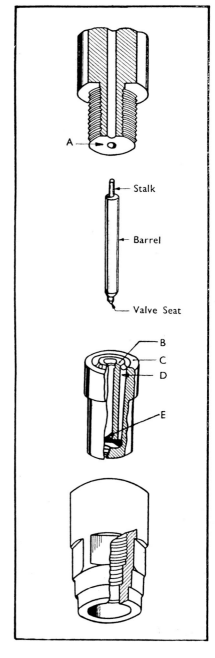

Injector nozzle components. See above for key to lettering and for cleaning instructions.

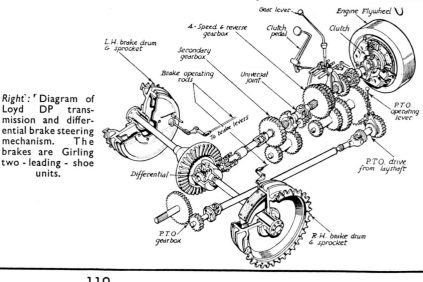

Right: Diagram of Loyd DP transmission and differential brake steering mechanism. The brakes are Girling two-leading-shoe units.

Crawler Tractor Scrapbook Pt 2 — Loyd

March, 1950 — BRITISH FARM MECHANIZATION

cross shaft contains two external grease nipples, which should be greased every 100 hours. Occasional adjustment of the foot pedal linkage is required to maintain an inch of free travel of the foot pedal before it meets clutch resistance. This adjustment is made by screwing, in the appropriate direction, the fork which connects the pedal rod to the clutch shaft. Screwing the fork on to the rod decreases the pedal free movement while the opposite movement increases it.

Transmission and Steering

From the engine clutch the drive is taken by a four-forward-and-one-reverse gearbox and a secondary reduction gearbox. The oil capacity of each gearbox is approximately 4 pints. Use a reputable brand of high-pressure gear oil, viscosity SAE140. Inspect the oil levels every 50 hours and maintain them to the height of each filler plug. Renew the oil every 500 hours, preferably after a few hours' continuous work.

From the reduction gear the drive is transmitted to a bevel gear differential and thence by two half axles to the track drive sprockets. Correct lubrication is the most important point in the maintenance of the differential. Use high-pressure gear oil of viscosity SAE140, and keep it to the level of the filler plug. Inspect the level every 50 hours and renew the oil every 500 hours.

Steering is by two hand-operated, two-leading-shoe Girling brakes, one of which controls the speed of each sprocket. Each brake can be adjusted by a square-headed threaded stud which protrudes from the brake carrier plate located behind the sprocket assembly. To adjust, turn the stud clockwise to its full extent, but do not use undue force. Then turn the stud anti-clockwise half a turn. This can be recognized by two clicks of the stud locking mechanism. Never try to adjust the brakes by altering the settings of the brake control rods. This would reduce braking efficiency.

Brake Lining Adjustment

When it is necessary to reline the brakes, adopt the following procedure. Split and remove a track chain from its sprocket. Slacken the brake adjuster stud. Remove the circle of eight nuts from the hub of the sprocket assembly and withdraw the axle shaft—the outer flange of which is normally retained by the eight studs. A large hexagon nut will be revealed, and should be removed by the special spanner supplied.

Next, withdraw an internally serrated lockwasher and oil-retainer assembly, and remove a second hexagon nut. Then withdraw the brake drum while taking care to prevent its fouling the exposed threads of the axle housing.

The two brake shoe assemblies will now be exposed, and can be removed and relined in the orthodox manner. The correct replacement lining and rivets are listed respectively as Don Lining B.S.5, part No. 32958B; and rivets, part No. GB 6110. These can be obtained from the tractor manufacturers or from Girling spares stockists.

Sprocket hub carrier bearings are lubricated via a nipple screwed into each hub. Give each nipple six shots from the grease gun every 50 hours.

Track Chains

Each track chain is fitted with bolt-on steel grouser plates, one of which is a half-plate. The half-plate is included so that a track chain can be shortened when the limit of track adjustment has been reached. At this stage, slacken the track adjuster bolt to its limit and remove the half-plate and its links. Do this to each track at the same time otherwise they will differ in length and, make the tractor veer out of line.

After removal of the half-plate, additional wear can be compensated by removing another pair of links and replacing the half-plate. When the limit of track life is reached renew each track and sprocket. It is false economy to renew a chain but not sprockets.

Correct track tension is important. To tighten a track, use the box spanner supplied with the tool kit and turn the adjuster bolt anti-clockwise until the distance from the top of the track frame to the underside of the chain is approximately 10½ ins.

Track Idler and Bottom Rollers

Each front track idler wheel is carried on roller bearings, which are lubricated via a grease nipple screwed into the idler hub. Give this nipple six shots of grease every eight hours. In extremely muddy conditions apply the grease gun until old grease exudes from the sides of the hub. Lubricate the top idler and bottom rollers likewise.

The bottom rollers are of steel and run on ball bearings carried on hardened steel axles. There are four bottom rollers to each track frame. The periphery of each rear roller is grooved to give clearance to the sprocket teeth: this allows the roller to operate as closely as possible to the sprocket, and thus maintain positive track alignment at this point. When ordering less than a complete set of bottom rollers, always state whether plain or grooved are required.

Rear view of left-hand-side steering brake back plate showing brake adjuster (A).

The track frames are mounted to the tractor at the rear on a transverse tube extending the width of the tractor in front of the sprockets. Phosphor-bronze bushes are located between the track frames and the tube. These bushes are lubricated via nipples, and should receive six shots from the grease gun every eight hours.

The front of each track frame is supported by a balance beam attached by a fulcrum pin under the front of the chassis. The balance beam permits vertical oscillation of each track assembly. The beam is mounted throughout on rubber bushes and requires no lubrication at any point.

The DP is manufactured by Vivian Loyd and Co. Ltd., Camberley, Surrey. Export Department: 1, Balfour Place, London, W.1.

Above: The track frame balance beam (A) requires no lubrication. This form of suspension allows vertical movement of each track frame.

Left: (A) dynamo drive belt adjuster; (B) fan shaft oil cup and (C) dynamo oil cup; (D) left- and right-hand decompressors.

The Loyd Dragon Tracklayer (contd.)

In addition to Ricardo precombustion chambers, the detachable cylinder head carries overhead valves, C.A.V. injectors, a valve decompressor and four starter fuse holders. The holders are provided to assist starting in ambient temperatures below 20 degrees F. The decompressor is for use when the engine is being started by hand.

As can be seen from Fig. 5, each fuse holder is essentially a pipe threaded into the head so that it projects into a precombustion chamber. Approximately 20 ft. of slow-burning fuse is supplied with each tractor. When required, a ½-in. length, lighted, should be inserted into each holder.

Fig. 3.—Right-hand final reduction gear housing with steering clutch and brake drum (A) and track sprocket hub (B).

The decompressor, which is hand-operated, is designed to hold both inlet and exhaust valves open to allow the crankshaft to be turned freely by hand. When sufficient speed has been reached, the decompressor is disengaged, after which the engine should start.

The inlet and exhaust valves are not interchangeable; each is fitted with two coil springs and the respective timing data is as follows:—

Inlet valve opens 7½ degrees before piston top dead centre, and closes 35 degrees after bottom dead centre. Exhaust valve opens 45 degrees before bottom dead centre, and closes 7½ degrees after top dead centre. Both inlet and exhaust tappet clearance is obtained by rocker arm adjustment and is 0.006-in. when cold. The firing order is 1-3-4-2 and the valve seat angle 45 degrees.

Fuel Injection System

The fuel injection system includes C.A.V. injector pump, diaphragm supply pump and two detachable element filters. Fuel is drawn from a 15-gallon capacity tank and forced through both filters by the diaphragm pump. The element of the first filter should be renewed at intervals, the length of which is determined by the quantity of dirt in the fuel—say between 100 and 500 hours. The second filter element requires no attention between engine overhauls.

The injector pump incorporates a pneumatic governor and excess fuel device. The latter is provided as an aid to starting: it is manually controlled by a push-button which reacts on the control rack and thus increases fuel supply until the engine has started, after which it automatically returns to the normal supply position.

The pneumatic governor, the main part of which is incorporated in the end of the injector pump, controls the engine speed from about 350 r.p.m. to a maximum of 1,600 r.p.m. and is itself subject to a hand lever by which any speed within this range can be pre-selected as a maximum. The control lever operates an inlet manifold butterfly valve which controls the amount of air allowed to pass into the engine.

All air entering the inlet manifold is cleaned by a Burgess oil-type air cleaner. This is mounted on the right-hand side of the dashboard, and offset so as not to interfere with the driver's normal furrow-wall line of vision.

Electrical Equipment

The standard electrical equipment comprises C.A.V. 12-volt dynamo type DBNB122, a C.A.V. 12-volt starter motor type BS512 K/14, and a 12-volt 113 ampère-hour capacity (10-hour discharge rate) battery. The dynamo is voltage controlled and the maximum charging rate is 13.3 amps. An ammeter is provided on the instrument panel.

Transmission

From the flywheel, engine torque is transmitted to a four-forward and one-reverse speed gearbox by a spring-loaded 11-in. single plate dry clutch, which is controlled by a hand lever. This is situated on the left-hand side of the driver, and has 9 ins. of free movement so

that at the moment of disengagement it is in its most convenient operating position. The clutch throw-out bearing is impregnated with graphite, consequently the clutch requires no routine lubrication.

The gearbox is bolted to the engine flywheel housing and contains sliding spur gears which give the following maximum speeds—against which are also given the manufacturer's drawbar pull figures:—

Gear	M.p.h.	lb. pull
1st	0.88	7,000
2nd	1.80	6,200
3rd	3.30	3,500
4th	5.50	1,900
Reverse	0.75	

From the gearbox, the drive is transmitted to a spiral bevel pinion and gear by two Layrub couplings in tandem. The spiral gear is bolted to a transverse shaft at each end of which is a multi-plate steering clutch arranged as shown in Fig. 7. These clutches transmit the

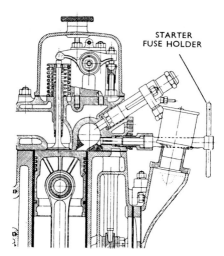

Fig. 5.—This drawing shows details of cylinder head, including combustion chamber, overhead valve, fuel injector and starter fuse holder.

drive to their respective track sprockets via spur reduction gears. The steering clutches are operated independently by hand levers, the correct free movement of which is 1½ ins. each. Contracting band steering brakes operate on the clutch drums and are controlled by two independently operated pedals. A hand control is also interconnected to a steering brake for parking.

There are four transmission lubrication reservoirs, the location and capacities of which are as follows: gearbox, 4 pints; bevel pinion and gear housing, 5 pints; final drive reduction gear housing (two), 2 pints each. A feature of the bevel pinion housing is that it has a combined oil filter and dipstick, the latter being marked in litres and pints.

Fig. 4 (left).—Rear of tractor showing steering brake adjusters (A); combined oil filler and dipstick (B); left-hand reduction gear housing drain plug (C); swing drawbar (D); right-hand reduction gear housing oil filler plug (E).

Fig. 6 (below).—Driver's view of controls, instrument panel and oil bath cleaner.

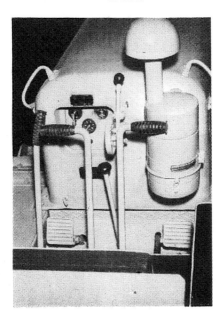

March, 1951 — FARM MECHANIZATION

The Loyd Dragon Diesel Tracklayer

Design and Operation

The Loyd Dragon is a new tracklayer, made by Vivian Loyd and Co., Ltd., Bridge Road, Camberley, Surrey. This tractor replaces previous Loyd models and is available with either a Turner or a Dorman-Ricardo engine. Both these power units develop 36.5 b.h.p.

THE Loyd Dragon is a Diesel-engined tracklayer with clutch and brake steering. It was introduced in 1950 to supersede all previous Loyd tractors, compared to which it shows, in our opinion, very considerable improvement in overall design. It is available with either a 4V95 Mark 2 Turner, or a 4 D.S. Dorman-Ricardo engine. The B.S.S. one-hour rating of each engine is 36.5 brake horse-power.

This article deals mainly with the Dorman-engined model, but an illustration and specification of the Turner engine are given on page 105.

The main features of the Dragon are identical on both Turner- and Dorman-engined models. These features include fabricated, welded ¼-in. section steel hull, hand-operated clutch, 4 forward and 1 reverse speed gearbox, multi-plate clutch and brake steering, rear drive sprockets, articulated track frame suspension, steel twin rail tracks with press fitted pins and bushes and detachable track shoes with integral grousers.

The total weight of the tractor is 8,200 lb. and its continuous drawbar pull is given by the manufacturers as 7,000 lb. in first gear (0.88 m.p.h.).

The cooling system is pump-assisted thermosyphon with a capacity of five gallons. Twin vee-belts drive a fan and water pump assembly mounted on the cylinder head. A C.A.V. 12-volt 160-watt dynamo is included in the belt circuit and is adjustable to obtain the correct belt tension. A thermostat is fitted to reduce engine-warming period. A heavy-duty detachable grid protects the radiator, the filler of which is funnel-shaped for easy filling.

Engine

The Dorman-Ricardo engine is a cold-starting, four-cylinder, four-stroke, over-head-valve unit incorporating Ricardo Comet II precombustion chambers. Total swept volume is 186.3 cubic ins.; compression ratio, 17.5 to 1; and maximum governed speed, 1,600 r.p.m.

The crankcase and cylinder block are formed by a one-piece casting containing renewable dry cylinder liners. The crankshaft has five main bearings, and main and big-end bearings are detachable white-metal-lined steel shells.

Aluminium-alloy pistons are standard and each carries three compression rings and one oil ring above the gudgeon pin, and a second oil ring on the skirt. Piston and connecting rod assemblies can be removed from the top of the cylinders.

Lubrication System

The lubrication system is pressure-fed by a gear type of oil pump driven by the camshaft. Oil is delivered direct by the pump to the main bearings and, via

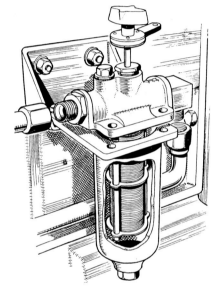

Fig. 2.—Cutaway view of the pressure-fed self-cleaning engine oil filter (see below).

drilled crankshaft, to the big-end bearings. The camshaft and overhead-valve gear is also pressure-lubricated, while the pistons and cylinders are lubricated by splash from the crankshaft. A pressure release valve set to raise at 40 lb. per square inch, is included in the circuit and the oil pressure is registered by a gauge mounted on an instrument panel.

A screen protects the pump inlet, and a manually operated pressure-fed self-cleaning oil filter is attached to the left-hand side of the crankcase. Details of this filter are shown in Fig. 2. To clean the element, the control seen on the top of the housing must be turned through one revolution; this rotates a laminated cleaner around the filter element and thus loosens dirt which then falls into a sludge trap contained in the base of the canister and provided with a sludge drain plug.

Sump oil capacity is 2¾ Imperial gallons, and the correct level is denoted by a dipstick. The oil viscosity recommendations, according to prevailing ambient temperatures, are as follows: above 90 degrees F., S.A.E. 40; between 32 degrees F. and 90 degrees F., S.A.E. 30; below 32 degrees F., S.A.E. 20. The oil should be renewed at 150 working-hour intervals. The self-cleaning filter control should be turned completely once every ten hours and the pump inlet screen removed and cleaned every 1,000 hours.

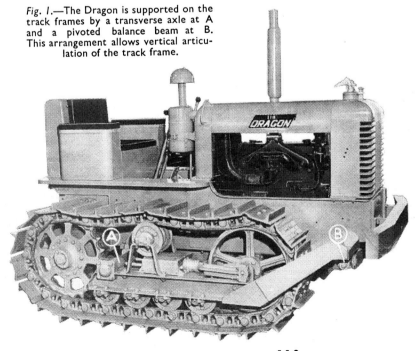

Fig. 1.—The Dragon is supported on the track frames by a transverse axle at A and a pivoted balance beam at B. This arrangement allows vertical articulation of the track frame.

March, 1951   FARM MECHANIZATION

Each track sprocket is bolted to a hub splined on to a taper axle carried by two Timken taper bearings. The standard track centre distance is 50 ins., but a 43-in. track centre can be supplied to special order. This narrow track model has a maximum track shoe width of 10 ins. and overall width of 59 ins.

The steel track links are fitted with interlocking bushes—that is to say, the bushes are slightly recessed into the outer links. The chief advantage claimed for this design is that the recessing affords protection against dirt which might otherwise enter the bush.

Twelve-inch-wide steel track shoes with integral grousers are standard equipment, but 14-in. shoes are available to special order. Ground pressure with 12-in. shoes is 6 lb. per sq. in.

The weight of the tractor is carried on the track frames at two points: at the rear by a transverse axle extending in front of both sprockets, and at the front by a balance beam pivoted on the hull under the radiator. As the beam is free to move about its fulcrum and because each frame can do likewise about the transverse axle, interdependent vertical oscillation of each track frame is permitted without sacrificing lateral stability.

There are four bottom rollers to each track frame; these are carried on bronze bushes and are interchangeable with the top idler rollers. Track adjustment is obtained by varying the position of the front idler, which is spring-loaded and free to recoil against shock loading.

Extra Equipment

A rear positioned power take-off and belt pulley is available as extra equipment. The six-spline coupling end of the power take-off shaft is 1⅜ ins. in diameter. Its maximum r.p.m. at governed engine speed is 550. Pulley diameter is 18 ins. and this gives a belt speed of 2,600 ft. per minute.

Fig. 7. Spiral bevel gear and multi-plate steering clutches.

Earth moving equipment available includes hydraulically operated angledozer to the following specification: blade width, 90 ins.; height, 28 ins.; maximum lift, 30 ins.; maximum dig. 9 ins.; angle 26 degrees in each direction. Attachment is by "A" frame anchored on transverse track frame shaft.

Manufactured by Vivian Loyd and Co., Ltd., Bridge Road, Camberley, Surrey, the standard model Dragon complete with either Dorman-Ricardo or Turner 4V95 Mark 2 engine is priced at £1,450 ex works.

ENGINES FOR THE LOYD DRAGON

Left : The Turner 4V95 Mark 2 engine showing valve covers (A), water pump (B), oil filler (C), lubricating oil pump junction (D) and starter motor (E).

SPECIFICATION :

Water-cooled, 4-cylinder, 4-stroke, overhead valve, 68 degree V form Diesel.

Bore, 3.75 ins.   Stroke, 4.5 ins.

Swept volume, 198.8 cu. ins.

Compression ratio, 17.5 to 1.

36.5 b.h.p. (one hour rating) at maximum governed r.p.m. of 1,600.

Right : The Dorman-Ricardo engine (4 D.S.) showing pneumatic governor control pipe (A), inlet manifold control lever (B), water pump drain pipe (C), dynamo (D), starter motor (E).

SPECIFICATION :

Water-cooled, 4-cylinder, 4-stroke, overhead valve Diesel.

Bore, 3.543 ins.   Stroke, 4.724 ins.

Swept volume, 186.3 cu. ins.

Compression ratio, 17.5 to 1.

36.5 b.h.p. (one hour rating) at maximum governed r.p.m. of 1,600.

A sectional drawing of the Loyd Dragon with Dorman engine appears overleaf.

KOMATSU

In 1917 Takeuchi Mining Co. (founded in 1884) established the Komatsu Ironworks to manufacture tools and mining equipment for their own use. In 1921 the Komatsu separated from the parent company to become Komatsu Ltd.

In 1931 Komatsu produced Japan's first crawler type farm tractor. 1947 saw the first D50 bulldozer, 1948 diesel engine production started and in 1969 Komatsu developed an amphibious bulldozer.

Komatsu's first venture outside Japan was in 1958 when a technical assistance agreement was signed with India to manufacture tractors. From there on Komatsu has ventured into many countries such as America, Singapore, Australia, Indonesia, & Europe. In 1975 the first Komatsu D50A was produced in Brazil and in 1989 the company began capital participation in Hanomag AG of Germany. In 1995 shipment began of D41 medium sized bulldozers manufactured by Komatsu Brasil to worldwide destinations.

The 1998 model line-up of Komatsu crawler tractors numbered 36. Komatsu bulldozers are manufactured in Japan, Brazil, Indonesia and Germany.

In 1991 Komatsu launched their new D575A-2 crawler, the world's biggest.
By 1996 10 of these machines were working in the United States, Australia and Japan. In 1994 the new D575A-2 Super Dozer was launched.
The D575A-2 Super Dozer was developed to meet the needs of companies requiring exceptional dozing performance where "cast Blasting" is a popular mining method.. The regular D575A-2 became the "Super Ripper". The D575A-2 Super Dozer had an operating weight of 147.9 tons (326,100lbs), a flywheel horsepower of 1,150hp (858kW), a 5,485mm (18') of track on the ground and a blade of 7,400mm (24'3") long by 3,050mm (18') wide with a capacity of 69 cub.metres (90 cu.yds)

Photo N S Komatsu, Sydney, Australia.

D575SD
FH 1150 HP (858 kW)
WT 150 tonnes (165.4 tons)

D575A
FH 1050 HP (784 kW)
WT 131.3 tonnes (144.8 tons)

D475A
FH 770 HP (574 kW)
WT 97 tonnes (106.9 tons)

D375A
FH 525 HP (391 kW)
WT 66 tonnes (72.8 tons)

Crawler Tractor Scrapbook — Komatsu

In 1985 the Komatsu 475A was introduced. Its Komatsu SA8V 170, eight cylinder diesel engine produced 770hp. It was followed by the 475A-2 of the same horsepower in 1989 and then the 475A-3 in 1996 with an increase to 860hp from a twelve cylinder engine.

The D475A-3 had an operating weight of 230,380lbs, 104,500kgs. Its height was 15ft, width 21ft 10inches, and length 34ft. Its torqflow transmission had three forward speeds from 2.2 to 6.8mph and 2.9 to 8.9mph in reverse.

A Komatsu sales illustration.

This Komatsu D375A was photographed on the Matahina, NZ, hydo dam in 1998. The dam was damaged in an earthquake and had to be rebuilt completely. The D375A was built between 1985 and 1991. It had a Komatsu-SA6D-170, six cylinder engine giving 508 ehp and weighed 123,140 lbs.

D355A
FH 410 HP (306 kW)
WT 53.2 tonnes (58.7 tons)

D275A
FH 405 HP (302 kW)
WT 48.8 tonnes (53.8 tons)

D155A
FH 320 HP (238 kW)
WT 41.3 tonnes (45.5 tons)

D155AX/D155A
FH 302 HP (225 kW)
WT 37.9 tonnes (41.8 tons)

Crawler Tractor Scrapbook Pt 2 — Komatsu

Komatsu D275s had a Komatsu S6D 170 six cylinder, diesel engine producing 405hp. Operating weight was approx. 111,840lbs, 50,720kg.
Height was 13ft, width 14ft 1 inch and length 29ft 10 inches.
Torqflow transmission gave three forward speeds from 2.4 to 7.3mph and reverse 3. to 9.3mph.

A Komatsu sales illustration.

Komatsu D150A and D155A had a flywheel horse power of 300 at 2000rpm. The 150A was a direct drive model with six forward and four reverse speeds and the D155A had Torqflow drive with three forward and three reverse speeds. The Komatsu S6D 155-4 was a diesel, turbo-charged, direct injection engine. Weight was 25,990kg (57,300lbs), length 5,395mm (214.4"), and height 3440mm (135.4").
This machine was photographed in Eagle Spares yard, Hamilton, NZ in 1998.

D85E	D85A	D85E-SS-2	D85E-SS-2A	D85P
FH 225 HP (168 kW)	FH 225 HP (168 kW)	FH 190 HP (142 kW)	FH 200 HP (149 kW)	FH 225 HP (168 kW)
WT 27.3 tonnes (30.1 tons)	WT 26.9 tonnes (29.7 tons)	WT 19.1 tonnes (21.1 tons)	WT 18.9 tonnes (20.8 tons)	WT 26.9 tonnes (29.7 tons)

Crawler Tractor Scrapbook Pt 2 — Komatsu

The Komatsu D85A-12 was built from 1968 to 1984. It had a Cummins-NH-220, six cylinder diesel engine giving 180ehp. Transmission type was Powershift. It weighed 47,650 lbs.

This machine was photographed in the ProTec yard, Rotorua, New Zealand in 1998. That's Rodney, the parts manager, giving an idea of the machine's size.

The Komatsu D75A was first made in 1970, powered by a 200hp, six cylinder diesel engine. The Torqflow transmission gave three forward speeds up to 6.5mph and three reverse to 8.3mph. Operating weight was approx. 30,930lbs.
A Komatsu sales illustration.

Komatsu D65EX-12 was built between 1994 and 1995. The Komatsu S6D125, six cylinder produced 190ehp. It had Torqflow transmission. Weight was 42,000lbs.

This machine was photographed in Eagle Spares yard, Hamilton, NZ in 1998.

D70LE
FH 180 HP (135 kW)
WT 17.1 tonnes (18.9 tons)

65EX/D65E-12
FH 180 HP (135 kW)
WT 18.3 tonnes (20.2 tons)

D65E-8
FH 165 HP (123 kW)
WT 17.7 tonnes (19.5 tons)

65EX/D65E-12
FH 180 HP (135 kW)
WT 18.3 tonnes (20.2 tons)

D65E-8
FH 165 HP (123 kW)
WT 17.7 tonnes (19.5 tons)

Crawler Tractor Scrapbook Pt 2 — Komatsu

Komatsu 58E were built between 1986 and 1991. Its Komatsu 6D125, six cylinder diesel engine produced 130ehp. weight was 23,920lbs. This machine was photographed at a Forestry Field Day at Rotorua in 1992.

Komatsu D50 was first produced in 1964 and went through various models to the D50A-18 of 1987. This D50A-15 was powered by a 90 hp, four cylinder diesel engine.
A Komatsu sales illustration.

This Komatsu 53A was busy filling in the local rubbish tip which had been closed, at Te Puke, NZ in 1999. The 53A were made between 1974 and 1987 and were powered by a Komatsu 4D 130-1 diesel engine giving 110ehp. Weight of the machine was 22,710lbs.

D60E
FH 165 HP (123 kW)
WT 16.7 tonnes (18.4 tons)

D58E
FH 130 HP (97 kW)
WT 14.9 tonnes (16.4 tons)

D53A
FH 124 HP (93 kW)
WT 13.4 tonnes (14.8 tons)

D50A
FH 120 HP (89 kW)
WT 12.8 tonnes (14.1 tons)

D41E
FH 105 HP (78 kW)
WT 10.5 tonnes (11.6 tons)

Crawler Tractor Scrapbook Pt 2

The Komatsu D45A was first made in 1975 and production appears to have continued, with variations, to 1985. They were powered by Komatsu S6D 105 six cylinder diesel engines giving 90hp. Three forward speeds were 2.1, 3.5, 5.8 mph and three reverse were 2.5, 6.9mph. Operating weight was 21.050lbs

Above: Colin Spence prepared a house site for the author in 1997. He was driving Komatsu D41A-3, serial #06138 which made it a 1982 model, they were built from 1981 to 1987. It was powered by a Komatsu 6D105-1, six cylinder diesel engine producing 90ehp at 2,350rpm. Three forward and three reverse speeds gave it 1.5 to 4.7mph.
Photo taken in 1997 at Rotorua.

The Komatsu D41E was first on show in NZ at the Fielddays at Hamilton in 1998. It had a 105hp, six cylinder Komatsu S6D102E diesel engine. Weight was 20,820lbs.

D50A
FH 120 HP (89 kW)
WT 12.8 tonnes (14.1 tons)

D41E
FH 105 HP (78 kW)
WT 10.5 tonnes (11.6 tons)

D41P
FH 105 HP (78 kW)
WT 40.9 tonnes (45.1 tons)

D37E
FH 80 HP (60 kW)
WT 6.7 tonnes (7.4 tons)

D37P
FH 75 HP (55 kW)
WT 7.5 tonnes (8.3 tons)

Crawler Tractor Scrapbook Pt 2 — Komatsu

This Komatsu 37E was photographed at a Forestry fieldday in Rotorua. They were made from 1986 to 1991 and had a 75hp Komatsu 4D 105-5 diesel engine..

Komatsu 37E-2 was on show at the National Fieldays. It was a hydroshift model with a Komatsu 6D 95L-1 diesel engine. Weight was 14,750lbs.

The Komatsu D33P-18 was powered by a Komatsu 4D95L 4 cylinder diesel engine, producing 70hp. Earlier models were rated at 63hp. It weighed 12,630lbs
This machine was photographed in Eagle Spares yard, Hamilton, NZ in 1999.

D31E	D31PLL	D31PL	D31P	D21A
FH 70 HP (52 kW)	FH 70 HP (52 kW)	FH 70 HP (52 kW)	FH 70 HP (52 kW)	FH 40 HP (29.5 kW)
WT 6.5 tonnes (7.2 tons)	WT 7.9 tonnes (8.7 tons)	WT 7.4 tonnes (8.2 tons)	WT 7.2 tonnes (7.9 tons)	WT 3.9 tonnes (4.3 tons)

Above: This calender scene was taken near Kaitaia, NZ in 1998. The Komatsu D31A-17 was made between 1981 and 1987. It had a Komatsu 4D105-3 63hp diesel engine. Torqflow gave it three speeds forward or reverse from 1.4 to 4 mph (2.2 to 6.5km/h). Weight was 13,650lbs, length 12'8" (3850mm), width 8'2" (2480mm), and height 8'7" (2620mm).

Below: Although it was built in 1966 this Komatsu D30-8, serial #2013, was not sold until 1971. The reason being that at that time Komatsu crawlers were not easy to sell and this one just sat and sat and sat in the dealer's yard. They were built from 1965 to 1972. It had four forward speeds of 1.6 to 5.5mph (2.5 to 8.9km/h) and two reverse of 2.1 and 3.9mph. Weight was 11,460lbs, length 136.6" (3470mm), width 99.6" (2530mm) and height 90.7" (2305mm).
This crawler belonged to Phillip Hawke when photographed in 1998. He bought it new.

Crawler Tractor Scrapbook Pt 2 — Komatsu

The Komatsu D21A-6 was powered by a Komatsu 4D95S-W, four cylinder diesel engine with a bore and stroke of 3.74 x 3.74" (95 x 95mm) giving 40 hp. It had a Torqflow transmission with two forward and two reverse speeds of 1.6 and 2.7mph forward (2.6 x 4.4km/h) and reverse 2.1 and 3.5mph (3.3 x 5.6km/h. Weight was 6,790lbs, length 10'8" (3250mm), width 7'1" (2170mm) and height 7'1" (2165mm). Photographed at the Te Puke rubbish tip filling site, NZ in 1998.

Komatsu D20A was made from 1965 to 1972 then was followed by the D20A-3 to 1983 and the D20A-5 to 1985. They were all powered by Komatsu 4D92, four cylinder, diesel engines giving 35hp.

Three forward speeds were 2.8 to 7.4km/h (1.7 to 4.6mph) and two reverse from 4.5 to 6.5km/h (2.8 to 4mph).

1986 the D20A-5 became the D20A-6 which was followed by the D20AG-7. Power was increased in 1990 with the 20A-6 being 40hp.

The Komatsu D20 PL-5 was made between 1978 and 1985 with an increase in engine power to 39hp. It weighed 8,330lbs.
This photograph was taken in Eagle Spares yard, Hamilton, NZ in 1999. The driver would not have much faith in the roll over bars!

D20PLL
FH 40 HP (29 kW)
WT 5 tonnes (5.5 tons)

D20PL
FH 40 HP (29 kW)
WT 4.4 tonnes (4.9 tons)

D20P
FH 40 HP (29 kW)
WT 4.3 tonnes (4.7 tons)

Crawler Tractor Scrapbook

Komatsu Underwater Bulldozer

Komatsu developed the world's first amphibious bulldozer, operable to a depth of 3 metres (10ft), in March 1969.

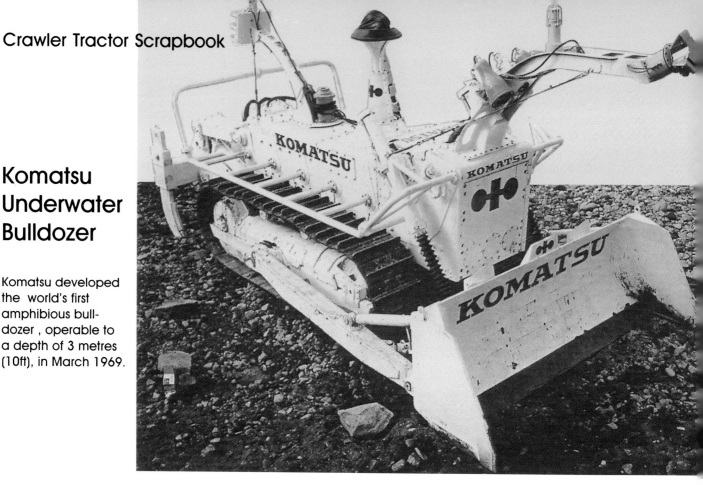

The Komatsu Underwater Bulldozer had a Cummins NRTO-6, 6 cylinder, diesel engine giving an output of 300hp at 1800rpm and a generator giving 170KVA/1800rpm. Operating weight on land was 74,960lbs(34,000kg) and in water 59,520lbs (27,000kg). Overall length was 330.7inches, width 147.6inches and height 129.1inches. Two forward speeds were 1.2mph (2.km/h) and 2.2mph (3.5km/h). Effective operating depth was 6.6 to 197ft (2 to 60m). Blade capacity 9.2 cub.yds (7m3). Standard equipment included a blade and ripper. The cable winch could hold 494ft (150m). It was equipped with transponder, compass, sonar, TV camera etc.

Illustrations from Komatsu sales literature

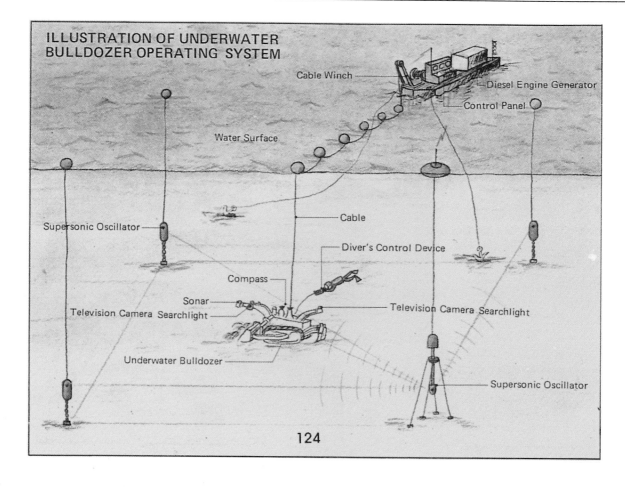

Crawler Tractor Scrapbook Pt 2 — Komatsu

The world's first amphibious bulldozer with remote radio control

The Komatsu amphibious bulldozer—the first of its kind in the world—is a revolutionary new product that opens up a whole new range of possibilities in the growing field of underwater exploration. Operable at depths down to 7 meters (23 ft), it is playing a leading role in underwater civil engineering and construction projects such as control and repair of river and harbor facilities, seashore maintenance, shore protection work and bridge construction and repair. A number of these units are already in operation throughout the world.

Operable Down to 7 meters (23 ft) depth
The maximum operating depth of seven meters greatly expands the scope of underwater operations. The vertical mast bends 90 degrees for safe travel under bridges.

Powerfully Built for Heavy Jobs
Large output and gross vehicle weight make possible heavy-duty excavation and earth moving, even in hard soil that can not be moved by a dredge pump or dredge bucket.

Blade Apron for Greater Efficiency
The dozer blade is equipped with an apron to prevent the loss of sand or soil due to underwater currents.

Easy Control and Maneuverability
Maneuverability is increased since no power unit vessel is used. Control levers and switches are easy to use with a minimum of skill and can be operated over long periods of time without fatigue. A choice of radio remote control or wire control lets the user select the one best suited to his job requirements.

Leakproof Construction
The torque converter, transmission, steering case and final drive housings are equipped with pressure equalizing devices to prevent leakage and protect sealed parts. The internal pressure is adjusted automatically to correspond to the water pressure outside. Engine vibration is also greatly reduced.

Built-In Safety Devices
When the bulldozer is submerged, warning lamps located on the top of the duct mast flash when the body is excessively inclined or when any abnormal condition or leakage occurs.

Lubrication-Free Levers and Links
The teflon bearings in the pin sections of the levers and links require no lubrication for the life of the machine.

Fast Parts Supply and Easy Maintenance
Watertight housings have inspection hatches for easy maintenance. Spare parts and parts adjustment methods are almost identical to standard bulldozers, thus they are easy to maintain or replace.

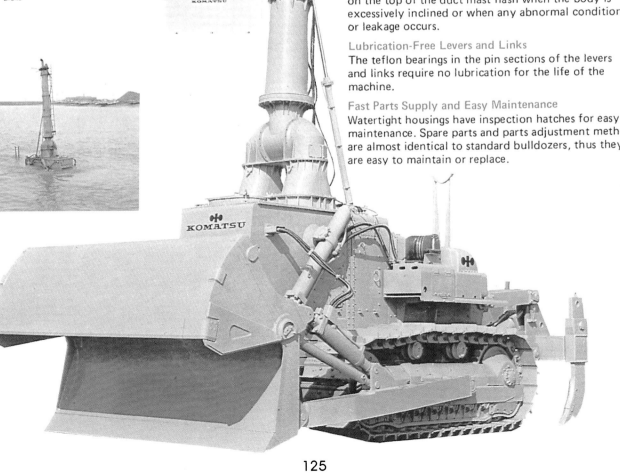

Crawler Tractor Scrapbook Pt 2 — Komatsu

KOMATSU

Model	Production Start - Finish	Comments A=standard, E=long frame, P=swamp, All even # models direct drive, uneven #s hydroshift, *=bare weight, hp=flywheel hp,				
D10A	1973 - 75	20hp,	2 cyl.	Kom.2D92-1	4160lbs*	
D20A	1965 - 72	35hp,	4 cyl.	Kom.4D92	7340lbs*	all D20s direct drive
D20A-2	1971 -	32hp,	4 cyl.	Isuzu DA220-QL	4,960lbs	
D20A-3	1973 - 83	35hp,	4 cyl.	Kom.4D92	7340lbs*	
D20A-5	1978 - 85	35hp,	4 cyl.	Kom.4D92	7,930lbs	
D20AD-5	1978 - 85	35hp,	4 cyl.	Kom.4D92		
D20A-6	1990 -	40hp,	4 cyl.	Kom.4D95S,	6,680lbs	
D20AG-7/PG-7	1992 -	40hp,	4 cyl.	Kom.4D92,	2,450/ 2,560lbs	also PG-7A
D20AP	1978 - 83	40hp,	4 cyl.	Kom.4D92		
D20P-2		35hp,	4 cyl.	Isuzu C221-PK	6,920lbs	
D20P-3		35hp,	4 cyl.	Kom.4D92	6750lbs*	
D20P-5/PL-5, P5A	1978 - 85	39hp,	4 cyl.	Kom.4D92	8,330lbs	
D20P-6,P-6A,PL6A	1986 -	40hp,	4 cyl.	Kom.4D95-S,	7,470lbs	
D20P-7/PL-7/PG-7	1992 -	40hp,	4 cyl.	Kom.4D95S,	7,430/ 9,020lbs	also PG-7A
D20PL/ PLL	1992 -	40hp,	4 cyl.	Kom.4D92		
D21A	1973 - 85	37hp,	4 cyl.	Kom.4D92	7,560lbs	all D21 hydroshift
D21A-3	1974 - 83	37hp,	4 cyl.	Kom.4D92	7560lbs*	
D21A-5	1986 - 95	39hp.	4 cyl.	Kom 4D94	7,980lbs	
D21A-6	1986 - 93	40hp,	4 cyl.	Kom.4D95S-W,	6,790lbs	
D21A-7/ AG-7	1993 -	40hp,	4 cyl.	Kom.4D95S,	6,750lbs	
D21AP-5	1986 - 95	39hp,	4 cyl.	Kom.4D95S,	6,750lbs	
D21E, D21E-6	1986 - 95	44hp,	4 cyl.	Kom.4D95S,	7,080lbs	
D21P-3	1974 - 78	37hp,	4 cyl.	Kom.4D92	6970lbs*	
D21P-5/PL-5	1978 - 85	39hp,	4 cyl.	Kom.4D94,	7,230lbs*	
D21P6	1986 - 93	44hp,	4 cyl.	Kom.4D95S-W,	7,580lbs	
D21PG-7		40hp,	4 cyl.	Kom.4D95S,	7,670lbs	
D30A-8	1965 - 72	52hp,	4 cyl.	Kom.DA220	11,460lbs	all D30 direct drive
D30A/P-12		55hp,	4 cyl.	Isuzu DA220-DC,	14990lbs	
D30A-15	1973 - 76	55hp,	4 cyl.	Kom.4D105-1	11,680lbs*	
D30P-15	1973 - 76	55hp,	4 cyl.	Kom.4D105-1	11,680lbs*	
D31A	197? - 87	63hp	4 cyl.	Kom.4D105-3,		all D31 hydroshift
D31A-15	1975 - 76	63hp,	4 cyl.	Kom.4D105-1	12,480lbs*	
D31A-16	1977 - 87	63hp,	4 cyl.	Kom.4D105-3	14,000lbs*	
D31A-17	1981 - 87	63hp,	4 cyl.	Kom.4D105-5,	13,650lbs	
D31A-18	1987 - 95	70hp,	4cyl.	Kom 4D95L		
D31A-20	1993 -	70hp,	4 cyl.	Kom.4D95L		
D31E	1986 - 91	63hp,	4 cyl.	Kom.4D105-5		
D31E-1	1983 - 86	63hp,	4 cyl.	Kom.4D105-3		
D31E-17	1986 - 91	70hp,	4 cyl.	Kom.4D105-5		
D31E-18	1987 - 95	70hp,	4cyl.	Kom.4D95L,	11,310lbs	
D31E-20	1993 -	70hp,	4 cyl.	Kom.4D95L,	11,310lbs	
D31P	1977 - 91	70hp,	4 cyl.	Kom.4D105-3		
D31P-15	1975 - 76	63hp,	4 cyl.	Kom.4D105-3	13,870lbs*	
D31P-16	1977 - 83	63hp,	4cyl.	Kom. 4D105-3,	14,880lbs	
D31P-17, PL-17	1981 - 87	66hp,	4 cyl.	Kom. 4D105-3,	12,570lbs	
D31P-17B	1986 - 88	70hp,	4 cyl.	Kom.4D105-3		
D31P-18/ 18A	1987 - 91	75hp,	4 cyl.	Kom.6D95-L,	12,630lbs	
D31P-20/PG-20	1993 -	70hp,	4 cyl.	Kom.6D95L,	11,770lbs	
D31PL/ PLL	1987 - 91	70hp,	4 cyl.	Kom.4D105-3		
D31PL-20/PLL-20	1992 -	70hp,	6 cyl.	Kom.6D95L,	13,540/14,590lbs	
D37E, D37E-1	1986 - 91	75hp,	4 cyl.	Kom.4D105-5		
D37E-2	1986 - 91	80hp,	6 cyl.	Kom.6D95L,	11,750lbs	
D37E-5	1993 -	80hp,	6 cyl.	Kom.6D95L,	11.770lbs	
D37P-2	1986 - 91	75hp,	6 cyl.	Kom.6D95L,	13,010lbs	
D37P-5	1993 -	80hp,	6 cyl.	Kom.6D95L	15,356lbs	

Crawler Tractor Scrapbook Pt 2 — Komatsu

Model	Years	Power	Engine	Weight	Notes
D40A/A-1	1975 - 83	80hp, 4 cyl.	Kom.S4D105	17590lbs*	all D40 direct drive
D40A-3	1981 - 87	80hp, 6 cyl.	Kom.6D105.	18,500lbs	
D40A-5	1987 -	95hp, 6 cyl.	Kom.6D105-1	23,320lbs	
D40P/P-1	1975 - 83	80hp, 4 cyl.	Kom.S4D105	19,510lbs*	
D40P-3, PL-3	1981 - 87	80hp, 4 cyl.	Kom.6D105,	20,040lbs	
D41A-A3	1981 - 87	90hp, 6 cyl.	Kom.6D105,	19,910lbs	
D41P, P-3	1981 - 87	90/95hp, 6 cyl.	Kom.6D105,	20,040	
D41P-5	1990 -	90hp, 6 cyl.	Kom.6D105		
D41E	1981 - 87	90hp, 6 cyl.	Kom.6D105,		
D41E-3	1981 - 87	95hp, 6 cyl.	Kom.6D105-1,	20,290lbs	
D41E-6/P-6	1996 -	105hp, 6 cyl.	Kom.S6D102E/ 6D105,	20,820lbs	
D45A	1975 - 77	90hp, 6 cyl.	Kom.S6D105	18,210lbs	all D45 Torqflow
D45A-1	1975 - 84	90hp, 6 cyl.	Kom.S6D105,	18,100lbs*	
D45P, D45P-1	1975 - 82	80hp, 6 cyl.	Kom.6D105,	20,020lbs	
D45P-3	1981 - 85	90hp, 6 cyl.	Kom.6D105		
D50	1964 - 73	86hp, 4 cyl.	Kom.4D120-10,	24,250lbs.	
D50A/P	1973 - 77	110hp 4 cyl.	Kom.4D130-1,	50A 22,050lbs, 50P 25680lbs	Direct drive
D50-11/D50P-11	1964 - 73	86hp, 4 cyl.	Kom.4D120-10,	24,250lbs	
D50A-15	1973 - 83	90hp, 4 cyl.	Kom.4D120-11,	24,250lbs	
D50A-16	1977 - 83	90hp, 4 cyl.	Kom.4D130-1	22,050lbs*	
D50A-17	1984 - 85	120hp, 6 cyl.	Kom.4D130-1/ 6D125,	22,510lbs	
D50A-18	1987 -	130hp, 6 cyl.	Kom. 6D125-1,		
D50P/PL-15	1974 - 83	90hp, 4 cyl.	Kom.4D120-11	24,290lbs*	
D50P/PL-16	1977 - 83	90hp, 4 cyl.	Kom.SL4D130-1,	25,680lbs	
D50P-17/ PL-17	1983 - 85	90hp, 4 cyl.	Kom.SL4D130-1	27,190lbs	
D50P-18, PA-18A	1987 -	130hp, 6 cyl.	Kom.6D125-1,	32,600lbs	
D53A	1974 - 87	110hp, 4 cyl.	Kom.4D130-1,	22,710lbs	
D53A-15	1974 - 77	110hp, 4 cyl.	Kom.S4D120-11	22,200lbs*	
D53A-16	1977 - 87	110hp, 4 cyl.	Kom.4D130-1	22,730lbs*	
D53A-17	1984 - 87	124hp, 6 cyl.	Kom.6D125,	23,740lbs	
D53P	1977 - 84	110hp, 4 cyl.	Kom.4D130-1,	26,230lbs	
D53P-16	1977 - 84	118hp, 4 cyl.	Kom.SL4d130-1,	26,230lbs	
D53P-17	1984 - 87	124hp, 6 cyl.	Kom.6D125,	27,630lbs	
D53P-18	1987 -	124hp, 6 cyl.	Kom.6D125-1		
D58E	1986 - 91	130hp, 6 cyl.	Kom.6D125,		
D58E-1	1986 - 94	130hp. 6 cyl.	Kom.6D125,	23,920lbs	
D58E-18	1988 - 91	130hp, 6 cyl.			
D58P-1	1986 - 91	130hp. 6 cyl.	Kom.6D125,	27,800lbs	
D60,	1964 - 76	130hp, 6 cyl.			
D60-3	1966 - 72	125hp, 6 cyl.	Cummins NHE-195-C1	30,670lbs	
D60A/P	1973 - 83	140hp, 6 cyl.	Cummins NG-220-C1		Direct Drive
D60A-6	1968 - 82	140hp, 6 cyl.	Cummins NG-220-C1,	33,620lbs	
D60A-7	1982 - 84	155hp, 6 cyl.	Cummins NG-220-C1.	28,570lbs	
D60A-8	1983 - 85	155hp, 6 cyl.	Cummins NG-220-C1	28,570lbs	
D60A-11	1986 -	160hp, 6 cyl.	Kom.6D125-1,	35,200lbs	
D60AG-6	1977 - 83	155hp, 6 cyl.	Cummins NG-220-C1		
D60E		165hp, 6 cyl.	Cummins NG-220-C1		
D60E-3	1966 - 72	140hp, 6 cyl.	Cummins NH-220-C1,	32,230lbs	
D60E-6	1968 - 71	155hp, 6 cyl.	Cummins NH-220-C1,	35,740lbs*	
D60E-7	1982 - 84	165hp, 6 cyl.	Cummins NH-220-C1,	29,740lbs	
D60E-8	1986 - 89	165hp, 6 cyl.	Cummins N855C,	29,700lbs	
D60F-7	1984	180hp, 6 cyl.	Cummins NTO-6	32,010lbs	
D60F-8	1985 - 88	172hp, 6 cyl.	Kom. S6D125,	33,069lbs	
D60PL	1973 - 77	140hp, 6 cyl.	Cummins NH 220-C1,	35,710lbs	
D60P-3	1966 - 72	140hp, 6 cyl.	Cummins NH-220-C1,	34,610lbs	
D60P/PL-6	1973 - 84	140hp, 6 cyl.	Cummins NH-220-C1,	31,970lbs*	
D60P-7, PL-7	1982 - 84	165hp, 6 cyl.	Cummins NH-220-C1	32,610lbs	
D60P-8/ PL-8	1984 - 88	165hp, 6 cyl.	Kom.6D125-1,	40,330/ 35,720lbs	
D60P-11	1987 -	190hp, 6 cyl.	Cummins NH220,	40,480lbs	
D60P-12	1992 -	190hp, 6 cyl.	Cummins NH220,	42,328lbs	

Crawler Tractor Scrapbook Pt 2 — Komatsu

Model	Years	Specifications
D63E, D63E-1	1986 - 95	140hp, 6 cyl. Cummins NH-220-C1, 25,550lbs
D63E-8	1986 - 89	140hp, 6 cyl. Cummins NH-220-C1
D65A	1973 - 83	140hp, 6cyl. Cummins NH-220-C1, 34,700lbs* *Torquflow*
D65A-6	1973 - 85	140hp, 6 cyl. Cummins NH-220-C1, 34,680lbs*
D65A-7	1982 - 84	155hp, 6 cyl. Cummins NH-220-C1, 29,010lbs
D65A-8	1984 - 85	155hp, 6cyl. Cummins NH-220-C1, 29,010lbs
D65E, D65E-1	1976 - 88	155hp, 6 cyl. Cummins NH-220-C1
D65E-6	1971 - 85	155hp, 6 cyl. Cummins NH-220-C1, 36,180lbs*
KBD65-6	1987 - 89	160hp, 6 cyl. Cummins NH-220-C1, 40,520lbs
D65E-7	1983 - 84	165hp, 6 cyl. Cummins NH-220-C1, 30,180lbs
D65E-8	1984 - 87	165hp, 6 cyl. Cummins N855C, 31,790lbs
D65E-12/EX-12	1994 - 95	180hp, 6 cyl. Cummins NH-220-C1/ Kom.S6D125-190, 32,780/ 32,890lbs
D65P	1976 - 87	140hp, 6 cyl. Cummins NH-220-C1
D65P-1	1976 - 83	140hp, 6 cyl. Cummins NH-220-C1
D65P-6	1976 - 83	155hp, 6 cyl. Cummins NH-220-C1, 33,400lbs*
D65P-7	1983 - 87	165hp, 6 cyl. Cummins NH-220-C1, 34,040lbs
D65P-8	1983 - 87	165hp, 6 cyl. Cummins NH-220-C1, 34,050lbs
D65P-12/PX-12	1994 - 95	190hp, 6 cyl. Cummins NH-220-C1/ Kom.S6D125, 35,800lbs
D65AR	1975 -	140hp, 6 cyl. Cummins NH-220-C1 radio controlled
DE65S-6	1975 -	270hp, 6 cyl. Kom.S6D155-4 amphibious in water up to 23ft, 7 metres.
D68E, D68E-1	1986 - 95	180hp, 6 cyl. Cummins S6D125, 33,990lbs
D68P, D68P-1	1987 - 95	180hp, 6 cyl. Cummins N855C, 36,210lbs
D70LE-12		180hp, 6cyl. Kom. S6D125, 41,840lbs, LE= Logging Emporer
D75A-1	1970 -	200hp, 6 cyl. Kom.S6D125, 30,930lbs
D80,	1964 - 80	180hp, 6 cyl.
D80A	1973 - 78	180hp, 6 cyl. Cummins
D80-8	1966?-	165hp, 6 cyl. Cummins NH-220-C1 41,230lbs
D80A-12	1973 - 88	180hp, 6 cyl. Cummins NH-220-C1, 38,800lbs*
D80A-18	1978 - 88	220hp, 6 cyl. Cummins NT-855, 51,180lbs
D80E-12	1973 - 88	180hp, 6 cyl. Cummins NH-220-C1, 50,040lbs*
D80E-18	1978 - 88	220hp, 6 cyl. Cummins NT-855 52,400lbs
D80F-18	1985 - 93	208hp, 6 cyl. Cummins NT-855, 48,700lbs
D80P	1975 - 76	180hp, 6 cyl. Cummins
D80P-18	1979 - 84	220hp, 6 cyl. Cummins NT855, 46,320lbs.
D83E-1	1987 - 95	205hp, 6 cyl. Kom.S6D125, 36,530lbs
D83P-1	1987 - 89	205hp, 6 cyl. Kom.S6D125, 39,090lbs
D85A	1974 - 84	180hp, 6 cyl. Cummins NH-220-C1, 47,200lbs*
D85A-12	1968 - 84	180hp, 6 cyl. Cummins NH-220-C1, 39460lbs*
D85A-18	1979 - 84	220hp, 6 cyl. Cummins NH-220-C1, 51,840lbs.
D85A-21/21B	1980 - 95	225hp, 6 cyl. Kom.S6D125, 40,200lbs/41,250lbs
D85E, D85E-18	1980 - 91	225hp, 6 cyl. Cummins NH-220-C1, 41,230lbs*
D85E-12	1974 - 81	180hp, 6 cyl. Cummins NH-220-C1, 50,710lbs*
D85E-21	1980 - 95	225hp, 6 cyl. Kom.S6D125, 42,040lbs
D85P	1974 - 84	225hp, 6 cyl. Kom.Cummins NH-220-C1
D85P-18	1979 - 88	220hp, 6 cyl. Cummins NT855, 46,980lbs.
D85P-21	1989 - 91	250hp, 6 cyl. Kom.S6D125, 48,060lbs
D85AR		180hp, 6 cyl. Cummins NH-220-C1, 47,200lbs, radio controlled.
D120,	1964 - 70	
D120-8		235hp, 6 cyl. Kom.6D155-2, 56,990lbs
D120-15		235hp, 6 cyl. Kom.6D155-2 58,750lbs
D120A-18		250hp, 6 cyl. Cummins NRTO-6-C1
D125	1964 - 70	
D125A-18	1968 - 72	250hp, 6 cyl. Cummins NRTO-66-C1
D125-18B		230hp, 6 cyl. Cummins NRTO-66-C1, 37,800lbs* first radio controlled.
D135A	1987 - 91	6 cyl. ?
D135A-1	1988 - 91	225hp, 6 cyl. Kom.S6D140,
D135A-2	1990 - 95	285hp, 6 cyl. Kom.S6D140, 53,920lbs
D150,	1970 - 72	?
D150A	1972 - 84	300hp, 6 cyl. Kom.S6D155-4

Crawler Tractor Scrapbook Pt 2   Komatsu

Model	Years	Specifications
D150A-1	1972 - 88	300hp, 6 cyl. Kom.S6D155-4, 60,160lbs*
D150A-1	1972 - 86	240hp, 6 cyl.
D150A-1	1986 - 88	300hp, 6 cyl.
D155	1970 - 72	??
D155A	1983 - 91	320hp, 6 cyl. Kom.S6D155-4, 32,060lbs*
D155AR		320hp, 6 cyl. Kom.S6D155-4, 32,0660lbs, radio controlled
D155A-1	1976 - 91	320hp, 6 cyl. Kom.S6D155-4 73,190lbs*
D155A-2	1988 - 95	320hp, 6 cyl. Kom.S6D140, 61,440lbs.
D155A-3	1994 -	302hp, 6 cyl. Kom.S6D140, 85,320lbs
D155AX/D155A	1994 -	302hp, 6 cyl. Kom.S6D155-4
D155AX-3 Super	1994 -	302hp, 6 cyl. Kom.S6D140, 86,420lbs, first Hydro,Mechanical,Transm.
D155AX-5	1999 -	310hp, 6 cyl, Kom.SA6D140E-1, 38.5 ton.
D155W		270hp, 6 cyl. Kom.S6D155-4, radio controlled, under water
D275A		405hp, 6 cyl. Kom.S6D170, 82,010lbs*
D275A-2	1991 -	405hp, 6 cyl. Kom.S6D170, 111,840lbs
D275A-3		405hp, 6 cyl. Kom.S6D170, 82,010lbs*
D355A	1970 - 81	410hp, 6 cyl. Kom.SA6D155-4
D355A-1	1971 - 91	410ph, 6 cyl. Kom.SA6D155-4, 79,590lbs*
D355A-3	1982 - 89	410hp, 6 cyl. Kom.SA6D155-4, 79,370lbs*
D355A-5	1989 - 93	410hp, 6 cyl. Kom.SA6D155-4, 80,800lbs.
D375A, D375A-1	1984 - 91	525hp. 6 cyl. Kom.SA6D170B, 98,680lbs.
D375A-2	1989 - 95	525hp. 6 cyl. Kom.S6D170, 101,190lbs
D375A-3	1995 -	525hp, 6 cyl. Kom.SA6D170E 145,590lbs
D455A	1975 - 83	620hp, ? cyl, Cummins VTA 1710-C800, 150,840lbs*
D455A-1	1975 - 88	650hp, 12 cyl. Cummins VTA 1710-C800, 154,900lbs.
D475A, D475-A1	1985 - 91	770hp, 8 cyl. Kom.SA8V 170, 148,400lbs.
D475A-2	1989 - 96	770hp, 12 cyl. Kom.SA12V140, 156,510lbs
D475A-3	1998 -	860hp, 12 cyl. Kom.SDA12V140, 230,380lbs
D575A-2 Super Ripper	1990 -	1050hp, 12 cyl. Kom.SA12V170, 212,740lbs, blade capacity 45.2 cub.m. (59 cub.yards)
D575A-2 Super Dozer	1995 -	1150hp, 326,100lbs, blade capacity 69 cub.m. (90 cub.yards)
G40	1936 -	50hp
TO-1	1943 - 44	88hp, ? cyl. weighed 11 tonnes. 80 built. The forerunner of the D50.
MK40,MK60, MK85, MK125, MK220 Komatsu badged Morooka, see Part 4.		

E & OE

Komatsu badged but Morooka built a MK125 rubber tracked crawler was demonstrated at the NZ National Fieldays in 1998. The Morooka crawlers are covered in Part Four.

JOHN DEERE

While John Deere started back in 1837 and Froelich's wheel tractor appeared in 1892 and they started getting serious about tractors in 1912, John Deere crawler tractors did not get a mention until the 1930s.

Lindeman crawler tracks were tried on a Model D in the 1930s but were not satisfactory. However there was some success with Lindeman tracks on the Model GPO (GP orchard) tractor in the early 1930s, approximately 25 being converted. The wheel tractor BO came on the market in 1936 and a Lindeman crawler version was made from 1943 to 1947.

The first John Deere designed crawler was the MC coming on the market in 1949. This was replaced with the 40C crawler in 1954 and the 420C appeared in 1955. The 430C crawler replaced the 420C in 1958. The same year saw the first of the John Deere industrial models painted yellow, the 440 crawler. Models 1010 and 2010 crawlers followed in 1960. 1965 introduced the JD350 and JD450 crawlers. The 1970s saw the B series arrive, the 350B and 450B. 1973 featured the 450C crawler. Next up was the JD550 and JD750 crawlers in 1975/76. The largest John Deere crawler appeared in 1978, the JD850. A new "G Series" came on stream in 1988 with the 400G, 450G and 650G.

Above: This 1946 John Deere BO with Lindeman tracks was on show at the Puget Sound Antique Tractor & Machinery Association's Threshing Bee in 1998 and belonged to John Corbin.
The JD BO had a two cylinder I-head engine with a 4 11/16 x 5½ inch bore and stroke giving 25.79 brakehp and 19.04dbhp at 1250rpm.

Left: Photographed at the Gore, New Zealand, rally in 1997 the John Deere MC crawler, serial number 14328, 1951, belonged to Mr I Barker.
The MC had a two cylinder engine rated at 1650rpm, with bore and stroke 4 x 4 inches. Dbhp was 13.88 and belthp 20.12. Weight was 4,293lbs. It had four forward speeds from .9 to 4.7mph.

Crawler Tractor Scrapbook Pt 2 — John Deere

JOHN DEERE Crawlers

Model	Production Start - Finish	Comments
BO Lindeman	1943 - 47	25bhp, 2cyl. JD petrol
MC	1949 - 52	20bhp, 18.2dbhp, 2 cyl. JD, 4,293lbs
40C	1953 - 55	20dbhp, 25.5bhp, 2 cyl. JD101-CID, 4,790lbs
350	1964 - 70	42ehp, 3 cyl. Perkins 3.152, 9,300lbs
350B diesel/petrol	1970 - 86	42ehp, 3 cyl. Perkins 3-152,/Deere 3-135, 8,220/ 8,115lbs
350C	1975 - 86	42ehp, 3 cyl. Deere 3-164, 10,099lbs
350D	1986 - 88	48ehp, 3 cyl. Deere 3-179D, 10,400lbs.
400G	1988 - 95	60ehp, 4 cyl. Deere 4-239D, 11,400lbs
400G LT	1994 - 95	60ehp, 4 cyl. Deere 4039D, 12,200lbs.
420C	1956 - 58	29ehp, 4 cyl. Deere, 5,010lbs
430C	1958 - 60	30ehp, 2 cyl. Deere, 5,159lbs.
440C	1958 - 61	30ehp, 25dbhp, 3 cyl. 6,919lbs
440IC	1958 -	30ehp, GM 30E, 7,103lbs, made in USA & JD Lanz
450 diesel/petrol	1964 - 70	57ehp, 4 cyl. Perkins 4-202/ Deere 4-180, 12,578lbs
450B	1970 - 76	65ehp, 4 cyl. Deere 4-219, 12,501lbs.
450C	1974 - 83	65ehp, 4 cyl. Deere 4-219T, 14,172lbs.
450D	1983 - 85	67ehp, 4 cyl. Deere 4-219T, 14,301lbs.
450E	1985 - 88	70ehp, 4 cyl. Deere 4-276D, 14,675lbs, also wide tracl 15,365lbs.
450E Long Track	1985 - 88	75ehp, 4 cyl. Deere 4-276D, 15,350.
450G	1987 - 95	70ehp, 4 cyl. Deere 4-276D, 15,266lbs, also LGP, long & wide track
550	1976 - 83	72ehp, 4 cyl. Deere 4-276T, 15,536lbs.
550A	1983 - 85	78ehp, 4 cyl. Deere 4-276T, 15,580lbs
550B	1985 - 88	78ehp, 4 cyl. Deere 4-276T, 15,501lbs, also long & wide track.
550G	1988 - 95	80ehp, 4 cyl. Deere 4-276T, 16,641lbs.
650G	1987 - 95	90ehp, 4 cyl. Deere 4-276T, 18,760lbs, also long & wide track.
750	1976 - 86	110ehp, 6 cyl. Deere 6-414T, 29,786lbs.
750B	1985 - 95	120ehp, 6 cyl. Deere 6-414T, 29,055lbs, also long & wide track.
750C	1996 -	140grosshp, 6 cyl. John Deere 6068T 33,736lbs.
850	1977 - 85	145ehp, 6 cyl. Deere 6-619T, 36,124lbs.
850B	1985 - 95	165ehp, 6 cyl. Deere 6-466A, 36,730lbs.
850C	1996 -	192grosshp, 6 cyl. John Deere 6081A, 40,155lbs.
1010C diesel /petrol	1960 - 65	42ehp, 4 cyl. Perkins 3.625, 6,439lbs 42ehp, 4 cyl. Deere, 6,337lbs.
2010C diesel /petrol	1960 - 65	52ehp, 4 cyl. Perkins 3.875, 8,864lbs 52ehp, 4 cyl. Deere , 8,762lbs.
8100T	1997 -	160ehp, 6cyl. Deere
8200T	1997 -	180ehp, 6 cyl. Deere
8300T	1997 -	200ehp, 6 cyl. Deere
8400T	1997 -	225ehp, 6 cyl. Deere
9300T	1998	Pre production model

E & OE

Crawler Tractor Scrapbook Pt 2 — John Deere

John Deere Send a Tracklayer

THE first John Deere tractor to be imported to this country for several years will be on the stand of H. Leverton and Co., Ltd. *(121)*. This is the latest Deere agricultural tracklayer, designated 440-IC, now being manufactured both in the U.S.A. and at the Lanz plant in Germany.

News that Leverton's intend to import this 35-h.p. model may surprise those who know them as *Caterpillar* concessionnaires. However, as Caterpillar are not likely to produce anything less powerful than the D4 (63 h.p.), the Deere machine is really complementary to the existing range handled by Leverton's.

The original American model was petrol-driven, and trials have been proceeding in Germany with various diesel units; it is understood that the fitting of a British Perkins diesel, with which the tractor will be shown, is by way of an experiment.

The 440-IC has four forward speeds and one reverse. Track design is based upon straightforward, well-tried principles with a simple single-point adjustment on each track. Frame, tracks and transmission have been designed to withstand such shocks as are encountered in loading, earthmoving and other industrial work. The price of the bare tractor, ex-depot, is £1,375.

FARM MECHANIZATION   December, 1958

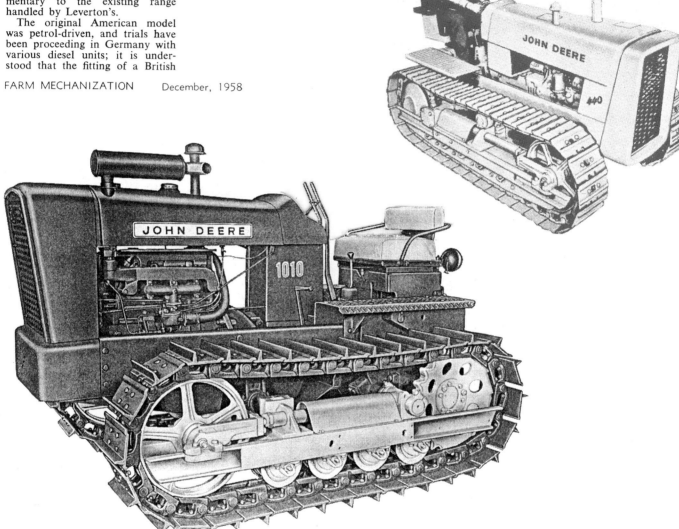

Both the John Deere 1010 and 2010 crawlers were on the market from 1960 to 1965.

The engine hp for the 1010 was 42 net at 2,500rpm, maximum dbhp 29.31 and maximum drawbar pull 7,484lbs at 1.36mph on 10inch tracks. The 2010 was 52 net. engine hp, 39.08dbhp and maximum drawbar pull 10,283lbs at 1.41mph on 12 inch tracks. Both were powered by John Deere engines.

The 1010 4 cylinder, diesel, engine had a bore and stroke of 3.625 x 3.5inches with a displacement 145 cub.in. The 2010 engine had a bore and stroke of 3.875 x 3.5 inches and a displacement 165cub.in.

Weight of the 1010 was 7,580 with extras, length 98½ inches, width 69¼ inches and height 53¼ inches. The 2010 weighed 9,645lbs with extras was 112½inches long, 64¼inches wide and 54 inches high to the top of the radiator.

1010 speeds were 1.41, 1.90, 3.30 and 6.5mph forward and reverse 1.9. 2010 speeds were 1.5, 1.8, 2.3, 2.8, 3.5, 4.3, 5.5, 6.7 mph forward and 1.7, 2.7, 4.1, 6.4mph reverse.

Steering was a clutch-brake system, multi-disk clutches and contracting band brakes.

Crawler Tractor Scrapbook Pt 2 — John Deere

Mr M Horrell had this 1950 John Deere 40C crawler on show at the Gore, New Zealand, rally in 1997.

The John Deere 40C had a two cylinder engine with a 4 inch bore and stroke giving 15.08 dbhp and 23.64 belthp at 1850rpm. Four forward speeds ranged from .82 to 5.31mph. Weight was 4,669lbs.

Above: This John Deere 420 was photographed in 1988 in New Zealand but I dont remember who owned it.

The 420C had a two cylinder engine with bore and stroke of 4¼ x 4 inches giving a dbhp of 18.65 and a belthp of 27.39 maximum. Four forward speeds were .87, 2.25, 3 and 5.25mph. Weight was 5,079lbs. It was available with four or five roller track.

Right: This John Deere 430C belonged to Bill Paterson, Dovedale, Nelson NZ when photographed in 1999. It was still in use. The 430C was basically the same at the 420C with a few more options and new styling.

Crawler Tractor Scrapbook Pt 2 — John Deere

The John Deere 350-C had a John Deere three cylinder diesel engine with a bore and stroke of 3.86 x 4.33inches giving a gross hp of 46 and net 42. It has 4 speeds forward of reverse from 1.4 to 6.5mph. It was produced from 1975 to 1986.

This crawler belonged to Bob Truman, Whakatane, NZ when photographed in 1999.

The John Deere 440 C was produced from 1958 to 1961in both USA and in Germany as John Deere Lanz. The original American model was powered by a petrol engine but the JDLanz had a 3 cylinder, diesel engine giving a maximum belthp of 30 and maximum drawbarhp of 25.
Forward speeds were 1.0, 1.62, 2.93, and 5.25mph and reverse 1.75mph. Height was 51", width 61", length 106¼" and weight 6,300lbs.
This machine belonged to Bill Higgs, Nelson NZ when photographed in 1999.

Both the John Deere 450E and the 650 G were photographed at the local dealer's yard in Lyndon, Washington State, USA in 1998. Some workmen wanted to know if I was a lawyer!
The 450E was built between 1985 and 1988. It was powered by a JD4-276D engine producing 70 engine horsepower. It had a powershift transmission and weighted 14,675lbs.

Crawler Tractor Scrapbook Pt 2 — John Deere

The John Deere 550 crawler was made between 1976 and 1983. It weighed 15,536lbs. Its engine was a JD 4-276T which produced 72 engine horsepower. Its transmission was three speed powershift.

The 650G weighed 20,400lbs. Its JD 4-276T turbocharged engine generated 90 engine hp. It was made from 1988 to 1995.
It had a 4 x 4 powershift transmission. Weight was 18,760lbs.
This machine was photographed at Lyndon, Washington State, USA in 1998.

The 750 B was built between 1985 and 1995. It had a J D 6-414T, 6 cyl. diesel engine giving 120ehp. It's weight was 29,055lbs and it was available in Low Ground Pressure, Long Track, Narrow Gauge and Wide Gauge models.

This machine was photographed in Carey and Riggs yard in Mapua, Nelson, NZ in 1999.

Crawler Tractor Scrapbook Pt 2 — John Deere

Another Carey and Riggs machine is this 850B, an early 850B model which has done 13-14000 hours work. Mark Carey is heading off to oversee the dam project they were building.

The 850 B was built from 1985 to 1995. It had a J D 6-466A six cylinder, diesel engine giving 145ehp. Weight was 36,730 for the standard model. The Long Track model weighed 38,601lbs.
Johnnie Gibbs is operating this machine for Carey & Riggs Ltd preparing sites and roads for logging.

The John Deere 850C came on the market in 1996 along with the 750C. They were described as the most powerful and productive crawlers that John Deere had produced to date.

The 850C was powered by a John Deere 6081A six cylinder 140 giving a 185 SAE net hp at 1,800rpm. The standard model (it was available in wide track and low ground pressure as well) weighed 40,155lbs, height 124inches, length 210 inches and blade width 128 inches.

This JD 850C was photographed outside the Cable Price Corp. premises in Rotorua, NZ.

Crawler Tractor Scrapbook Pt 2 — John Deere

The John Deere 8400T was first displayed in NZ at the National Fieldays in 1998. Production started in 1997. It was powered by a JD 6 cylinder diesel engine producing 225 ptohp. Weight was 17,030lbs. Speeds were 16 forward and 4 reverse.

8400T
225 (168 kW)
2,200 rpm
6-cylinder, in-line, wet-sleeve, valve in head
Turbocharged and aftercooled
8.1 L (496 cu. in.)
4.56x5.06 in. (115.8x128.5 mm)
In-line pump, electronic governor
130 U.S. gal., 108 Imp. gal. (492 L)
140 amps/1,850 cca (two, 12-volt batteries)
Outboard planetary
60 to 88 in. (1524 to 2235 mm)
16 in. (406 mm); 24 in. (610 mm)
8.3 psi (57.56 kPa)
5.7 psi (39.53 kPa)
1¾ in. (44.5 mm), 1,000 rpm
Pressure-flow compensated
3 (4th factory installed, 5th field installed)
30 U.S. gpm, 25 Imp. gpm, (1.9 L/s)
Hydrostatic differential
Cat. 3, 15,650 lb (7099 kg)
16-speed forward, 4-reverse Power Shift

This "Field Test Unit" John Deere 9300 was on display at Farm Progross in Windfall, Indiana in September 1998.
Photograph from Martin Smits collection

Crawler Tractor Scrapbook Pt 2

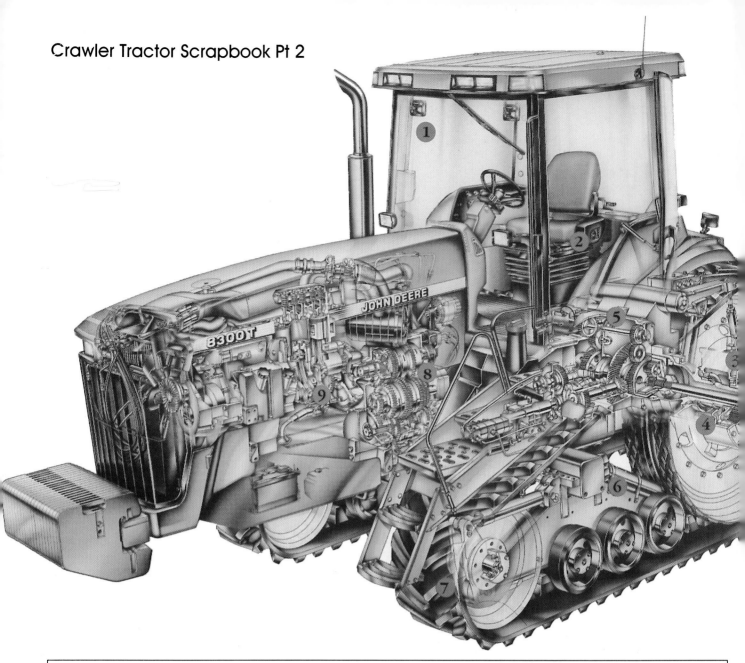

1. Exclusive Command View™ cab offers unprecedented visibilty- 62 square feet of tinted glass.
2. Exclusive Personal-Posture® seat with aircushioned comfort and exclusive ride-zone protection keeps you comfortable all day long.
3. Exclusive tread adjustment lets you change tread settings in half the time it takes to change tread width on competitive models.
4. The heavy duty outboard final drives, cast hub and strong rear axle handle high torque loads easily.
5. Exclusive "speed sensitive" variable-rate differential steering improves handling in the field and on the road.
6. High-pressure accumulator ensures proper tread tension.
7. Choose either 16- or 24 inch track widths.
8. The ultra-efficient 16 speed Power Shift transmission maximizes power flow and eases operation.
9. Record-setting 8.1 L engines are standard on all four models. You get up to a 10 percent power bulge and can maintain high-torque performance to as low as 1,000rpm.

From a John Deere Sales Brochure *Sign of the Right Choice*

LANZ

During the 1914-1918 war Lanz produced at least one crawler for military purposes. After success with half tracks on Bulldog tractors in the 1920s it was decide in 1932 to build a full crawler. After much experimenting and many trials with the likes of the HR6, using as a base the HR8, a new Crawler Bulldog of 38hp came on the market in 1935. A change in measuring hp came about in 1936. "Normal continuous output" was changed to "maximum output over one hour" thus the 38 hp Crawler Bulldog became 45hp. Some export models were designated 44hp.

1937 saw an increase in power to 55hp and the Model became the HRK.

Production ceased in 1946 with a little over 4,000 crawler Bulldogs having been manufactured, the peak year being 1943.

Photographs: This Lanz Model T, D1550, belonged to Mr R A J Copland when photographed in 1992 at a rally in Rangiora, New Zealand. The single cylinder engine produced 55hp. It was made in 1938.

Lanz

Model	Production Start - Finish	Comments
Army	1914 - 18	
Pre-series	1933 - 34	30/38hp, one cyl.
D6550 HR6 base	1934	38hp, one cyl.
D9550 HR8	1935 -	38hp "normal continuous output, 1936- 44/45hp maximum output.
D9551		same with electric lights, ignition & horn.
D9552		same extras without electric starting.
		Export models 44hp.
D1550 HRK 55 Model T	1937 - 46	55hp, one cyl.
D1560 HRK55	1940 - 46	55hp, one cyl.
D1561 HRK55	1942 - 46	55hp, single cyl. Including electrics,
D1571 HRK55	1941 - 44	55hp, one cyl. Designed for road construction, 12 speed gearbox.
D1581 HRK55	1941 - 44	55hp,? one cyl. Designed for earth moving etc.
		Crawler tractor production ceased in 1946.

E & OE

Crawler Tractor Scrapbook

This Lanz T crawler belonged to Max Hogg when photographed at the Henty rally, Australia, in 1999. It was a 1937 HRK D9550 model with 55hp. Prior to 1937 export models were 44hp.
Its single cylinder engine could be fueled by gas oil, coal tar-oil, diesel oil, paraffin oil, plant oil etc. Its bore and stroke were 225mm x 260mm and revolutions 750rpm. Weight was approx. 5000kg, length 3.070m, width 1.960m and height 2.380m. Speeds forward were 3.2, 3.9, 4.8, 5.5, 6.6, 8km/h and reverse 4.2 and 5km/h.

Crawler Tractor Scrapbook Pt 2 — Vickers

Vickers

In 1950 Vickers-Armstrong of the United Kingdom announced they were going to produce a crawler tractor. It was not until March 1952 that production actually started for the Vickers VR.180. It had a nine year production life. The first VR.180 was delivered on 31/3/52 and by 31st December 1959 889 Vigors and 21 VR110 Vikons had been built. Only a very few Vigors were made in 1960, if any, as for the last few months only spare parts were made.

Local contractor opinion was that the VR.180 had some wonderful design features. If only the Vickers-Armstrong engineers had come out and discussed problems, as they were invited to do, it could have been a success. However as it was the VR.180, while handling towed implements such as scoops was good, bulldozing was a disaster. As soon as pressure was placed on the dozer blade the tracks lifted and traction was lost.

Only two models were made, the first became known as the Vickers Vigor, the second one was the Vickers Vikon.

In April 1961 production ceased.

> A Vickers VR.180, "Vigor", is put through its paces in Northland, New Zealand in 1956.
> Graeme Craw Collection

The "VR.180" Tractor Arrives

Export Markets will Absorb much of Early Production

News which will, no doubt, be received with special interest in industrial and constructional engineering circles throughout the world is that production of the British "Vickers VR.180" tractor begins at the end of this month. It was in August, 1950 (p. 618), and the following October (p. 927), that we first gave details of this imposing heavy track-type machine, and although, as we emphasised at the time, it is not intended for agricultural use, the tractor has roused a curiosity spreading far beyond the spheres for which it is expressly designed.

Various items of advance information concerning the "VR.180," the occasional glimpses of pre-production models at some of the major shows, and the odd scraps of detail gleaned in connection with the long and rigorous series of field trials made, have all contributed to the mood of high expectancy that now prevails, but, if possible, a new brochure just released "whets the appetite" even more.

In some 40 admirably-arranged coloured pages, it provides not only a specification of the tractor, but full descriptions of the many absorbing features incorporated. The impressive "Rolls-Royce" engine, the articulated suspension, track and track wheels, transmission, gearbox, cooling system, unit construction and controls all receive close attention, while the reader's appreciation of certain points is much assisted by sectional illustrations. The influence of design on operational handling, the range of equipment available, and the essential supporting sales and service organisation are also discussed.

World distribution of the "VR.180" is, of course, in the hands of Messrs. Jack Olding & Co., Ltd., Hatfield, Herts., who advise us that the larger part of production is destined for export markets. It is believed, with some justification, that the machine will both help to satisfy world demands for better heavy earth-moving equipment and, at the same time, add to the prestige of Britain's engineering capacity generally. The tractor and its ancillary units are to be shown by Messrs. Olding at the forthcoming British Industries Fair, at Birmingham.

Farm Implement & Machinery Review — March 1952

Crawler Tractor Scrapbook Pt 2 — Vickers

A Vickers Vigor is carefully unpacked in Wellington, New Zealand, after its arrival by sea from England.

Photograph from Graham Blackley's Collection

Not so careful was the tying down of this Vickers Vigor onto a truck. Someone would have had their fingers rapped over this mess. Apparently the machine was written off.
Photo credit The Dominion Newspaper.

A new Vickers Vigor ready for work with an Onions Model 12-16 Scraper. The Vickers crawlers were ideal for this type of work, ie pulling scrapers.
Photograph from Graeme Craw's Collection

VICKERS FRONT CONTROL UNIT
VFCU · 180

GENERAL DESCRIPTION
Front-mounted single drum winch driven through a two-stage reduction gear connected by a flexible coupling direct to the crankshaft. For use with Vickers Universal Dozer Blade

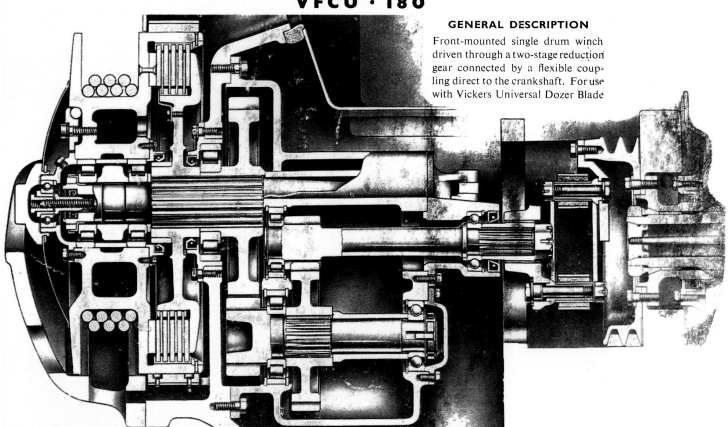

Line Speed (at engine governed speed of 1800rpm) Bare Drum 440ft per minute..134.1 metres p.m. Full Drum 556 ft p.m..169.4 m.p.m.
Cable Size 1/2 in. diameter (6 x 19)..12.7.. **Clutch** Multi plate with forced ventilation. **Brake** Automatic self wrapping band.
Control Single lever on RH side of driver's seat. **Weight** 570 lb. .. 258 kg
Drum Diameter.. 9 in... 228.6mm. Length 2.5 in .. 63.5mm Capacity 94 ft .. 28.65metres

Crawler Tractor Scrapbook Pt 2 — Vickers

Photograph from Graham Blackley's Collection

Above: Initially the Vickers Vigor crawler tractors were brought into New Zealnd by the Government. It is understood they were used around the Wellington area including work on the Wellington Airport and this housing estate at Porirua. It looks like there is another Vickers Vigor in the background. The total number brought into the country is not known exactly but is thought to be about 25.

For some reason the NZ Government was keen to get its departments to take on Vickers machines. Photographs taken by Jim Spiers in the 1950s show a demonstration in the forest in the Bay of Plenty area. The chaps were not impressed and suggested a tug of war with the Caterpillar D8. "The D8 pulled its pants off!"

Crawler Tractor Scrapbook Pt 2 — Vickers

June, 1952 FARM MECHANIZATION

The VR-180 with cable operated bulldozer which is also designed and manufactured by Vickers-Armstrongs Ltd.

The Vickers VR-180

The VR-180 is a worthy challenger to the virtual monopoly held by large American tracklayers. It constitutes a complete breakaway from the conventional and is capable of moving earth at high speed

THE VR-180 is a compact, powerful and unorthodox tracklayer designed and manufactured by Vickers-Armstrongs for earth moving and allied duties. Weighing 32,500 lb., it incorporates a Rolls-Royce oil engine of 180 belt horsepower and has a top speed of approximately 10 miles per hour.

Since prototypes were first tested in 1949, pre-production models have been used on open-cast coal sites and civil engineering projects at home and abroad. Quantity production began during March at the Scotswood Works, Newcastle, of Vickers-Armstrongs Ltd.

Priced at £7,475 ex works, the VR-180, together with earth-moving equipment, which has been developed concurrently, is being distributed throughout the world by Jack Olding and Co. Ltd., Hatfield, Herts. This company have already selected dealers in principal territories including Australia, India, Pakistan, New Zealand and Central and Southern Africa, and a condition of appointment is that the dealer must maintain equipment and personnel capable of providing an efficient after-sales service for the supply of spare parts and repair of the VR-180.

A sectioned, perspective drawing of this tractor appears on pages 234 and 235 and a detailed specification is given on page 233. In addition to the engine, the main features include hull construction, six-forward and three-reverse speed gearbox, clutch and brake steering, three-point articulated suspension, and reflexing, sealed lubricated tracks.

The main component in the front half of the tractor is the hull. This is fabricated and cross-braced from ⅜ in. steel plate. It supports the engine and its accessories and is mounted on the forward track wheel beam at the front and attached to the main transmission housing at the rear. The rear connection is made by fourteen ¾-in. bolts and it is claimed that these can be removed and the tractor split into two basic assemblies within two hours.

In addition to containing the engine and thus protecting the sump and other vulnerable components, the hull also provides 2-in. thick anchorage points for bulldozer attachments.

The Rolls-Royce oil engine is a six-cylinder, supercharged, four stroke, which develops 180 belt horse-power at its maximum governed speed of 1,800 r.p.m. A single casting forms the cylinder block and crankcase. It incorporates wet cylinder liners and a seven-main-bearing nitride-hardened crankshaft. Aluminium pistons are attached to forged connecting rods by fully floating gudgeon pins.

There are two cylinder heads. Each covers three cylinders and includes valves, hardened valve seat inserts and C.A.V. injectors. The camshaft, injector pump and oil pump are gear driven.

Lubrication is by a dry sump system which is designed to provide adequate lubrication irrespective of the angle of operation. The system includes an

(Continued overleaf.)

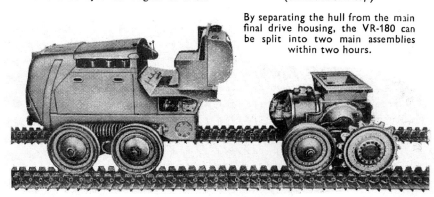

By separating the hull from the main final drive housing, the VR-180 can be split into two main assemblies within two hours.

Crawler Tractor Scrapbook Pt 2 — Vickers

The Vickers VR-180 (contd.)

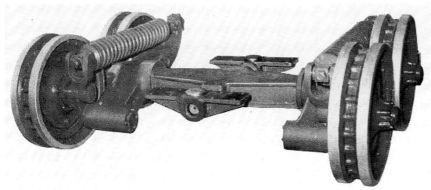

Articulated front wheel and track beam assembly which supports the front of the tractor beneath the hull.

Exploded view of track link assembly. Lubricant is hermetically sealed within the bushes during manufacture and requires no attention between major overhauls.

oil-to-water heat exchanger, one pressure feed pump, two scavenging pumps and three full-flow filters.

From the pressure pump, oil is forced through the heat exchanger and filters to a gallery connected to the main bearings.

The crankshaft and connecting rods are drilled to provide a full-pressure supply of oil to the big ends and gudgeon pins. A metered, low-pressure supply is taken from the main gallery to the supercharger, camshaft, valve rockers and ancillary components.

The heat exchanger attached to the left hand side of the sump is an important factor in the design of the engine. Coolant is circulated around pipes which carry the lubricating oil, so that the oil is warmed on starting and cooled during normal running.

In addition to the heat exchanger, the cooling system includes pressurized radiator, fan, centrifugal pump and thermostat. Twin vee-belts drive the fan from a pulley which is mounted on a viscous oil-type torsional damper situated on the front of the crankshaft.

A second pair of vee-belts drive the pump from a pulley which is directly coupled to the timing gears. This pulley is slotted to accommodate a tommy bar so that the engine can be turned by hand during repairs and adjustment. This provision is one of many that have been made to facilitate maintenance.

The air drawn through the radiator core first passes through horizontal screens fitted in the top of the tractor. This position provides the cleanest air that is likely to be available during work; furthermore, the gauge of the screens is finer than the air passages through the core, consequently the core is protected from dirt.

The radiator pressure valve is set to release at 7 lb. p.s.i. This pressure raises the boiling point of the coolant above normal and thus enables the engine to work in high temperatures and at high altitudes without overheating. When the engine is warming up, the thermostat restricts the circulation of the coolant to the engine and heat exchanger.

Fuel and Air System

The fuel and air system includes three oil bath air-cleaners, a supercharger, fuel filters and C.A.V. injection equipment. From a sixty-gallon tank, a mechanical pump draws the fuel through three filters before passing it to the injector pump.

A variable speed centrifugal governor is contained in the injector pump. It is interconnected to hand and foot controls, and limits the engine speed to any range within that pre-selected by the manual control.

The engine starts from cold and an excess fuel device is incorporated in the fuel pump to facilitate starting when the ambient temperature is low.

Electrical System

Standard electrical equipment comprises four 6-volt, 185 ampere-hour batteries connected in series; a $5\frac{1}{4}$ in. diameter voltage controlled dynamo; 24-volt axial type starter motor, and lighting set. As an example of detail, it is worth noting that all the electrical cables, excepting those to the lamps, are housed in armoured and sealed conduits. Furthermore, a socket and cable is provided so that additional batteries can be connected if the tractor batteries become exhausted.

Transmission

An 18-in., single-plate, foot-operated dry clutch transmits the drive from the flywheel to the gearbox through a flexible rubber coupling.

Constant mesh, helical gears provide six forward and three reverse ratios. Gear changes can be made while the tractor is moving and only three movements of the levers are required to change from a bottom gear of 2.18 m.p.h. to top which gives 9.73 m.p.h.

A spiral bevel pinion and gear transmit the drive from the gearbox, through multi-plate steering clutches, to the track drive sprockets. Each clutch contains eight spring-loaded friction discs that are serrated on the outer edges.

Steering is by two independent levers which actuate hydraulic servo mechanism for clutch disengagement. An external contracting band brake acts on each clutch drum and is applied by a linkage which is connected to the clutch release levers and which cannot be actuated until a clutch is fully released.

The hydraulic servo pump, which operates the clutch control mechanism, is driven from the gearbox and it is impossible for the clutches to be disengaged when the engine is not running. An overriding brake pedal is provided for parking.

Undoubtedly the most outstanding features of the VR-180 are its suspension and track. Fully articulated track wheels and reflexing tracks enable the tractor to negotiate rough terrain smoothly and at high speed.

Four large diameter wheels, coupled in pairs, are contained within each track circuit. The rear track wheel on each side also incorporates the driving sprocket. The driving sprocket and track wheel rims are insulated from their respective hubs by rubber rings, as shown in the main drawing on pages 234 and 235. These rings allow deflection

The 3-point fully articulated suspension enables the track to flex in both directions.

Crawler Tractor Scrapbook Pt 2 — Vickers

June, 1952   FARM MECHANIZATION

under load and isolate the tractor from track vibrations.

The rear sprocket and wheel assembly incorporates the final drive gear train in a housing mounted on a central pivot. This arrangement allows sprocket and wheel to oscillate radially and thus conform to undulating ground. The front pair of wheels can move radially on their supports and laterally on a central pivot which attaches the front beam to the hull. Each front wheel axle is connected to a compression spring which allows the track to adjust its own tension according to the degree of deflection demanded by varying ground levels. The track can flex in both directions and thus maintain a high degree of ground contact over irregular terrain.

The track pins and bushes are lubricated by oil which is sealed in the bushes during manufacture and which requires no attention until a major overhaul of the track is necessary.

Earth moving equipment designed and manufactured by Vickers-Armstrongs for the VR-180 includes a cable dozer, scraper, and cable control units. Specifications of tractor and equipment appear below.

Right-hand side of Rolls-Royce oil engine. The opposite side of this engine, including supercharger, can be seen in the drawing on pages 234 and 235.

SPECIFICATION OF VICKERS VR-180 TRACTOR AND ANCILLARY EQUIPMENT

The Vickers VR-180

Engine model: C6SFL Rolls-Royce supercharged oil engine. Four-cycle, liquid-cooled, direct-injection, six-cylinder (vertical in-line). Bore, 5.125 ins.; stroke, 6.0 ins.; swept volume, 742.64 cubic ins.; 180 belt horse-power. All-speed governor set to limit speed under load to 1,800 r.p.m. Drawbar h.p., 150.

Main features: Integral cast-iron crankcase and cylinder block. Wet cylinder liners. Dry sump, incorporating oil and water heat exchanger. Full-flow oil filters. External viscous-oil-type crankshaft damper. Twin belt pulley for driving the radiator cooling fan, and 60 h.p. power take-off from front end of crankshaft.

Engine accessories: Electrical starter. Dynamo. Engine service counter. Positive-displacement-type supercharger. Air cleaners. Triple fuel filters.

Starting method: Electrical axial-engagement starter, 6 ins. dia., 24-volt; dynamo, 5½ ins. diameter, 24-volt; batteries, four 6-volt 185 ampere-hrs.

Drawbar pulls: The following are drawbar pulls at rated engine speed observed during manufacturer's tests: First, 26,100 lb.; second, 17,000 lb.; third, 14,500 lb.; fourth, 8,240 lb.; fifth, 6,720 lb.; sixth, 3,140 lb.

At maximum engine torque greater pulling power is developed and the following are calculated values for maximum drawbar pulls at reduced travel speeds: First, 29,500 lb.; second, 19,250 lb.; third, 16,400 lb.; fourth, 9,300 lb.; fifth, 7,600 lb.; sixth, 3,600 lb.

Travel speeds at rated engine speed: First, 2.18 m.p.h.; second, 3.25 m.p.h.; third, 3.76 m.p.h.; fourth, 5.60 m.p.h.; fifth, 6.53 m.p.h.; sixth, 9.73 m.p.h. Reverse, 2.66 m.p.h.; reverse, 4.58 m.p.h.; reverse, 7.98 m.p.h.

Dimensions: Gauge (centre to centre tracks), 80 ins.; length of track on ground (centre drive sprockets to centre front wheel), 104 ins.; area ground contact, 4,992 sq. ins. (with standard track shoes); overall length, 176¾ ins.; overall height, 86⅜ ins. (tip of grouser to highest point); width, 104 ins.; ground clearance (from lower face of standard track shoe), 10 ins.; height of drawbar above ground (from lower face of standard track shoe), 17¾ ins.; lateral movement of drawbar (measured at drawbar pin), 35½ ins.; width of standard track shoe, 24 ins.; height of grouser (from upper face of standard track shoe), 1¾ ins.; diameter of track-shoe bolts, ⅝ in.; track pins, 1⅝ ins.; track-pin bushings, 2¼ ins.

Steering: Type, clutch and brake; clutch friction material, bonded asbestos; number of friction surfaces each clutch, 16; type of clutch release, hydraulic. Brakes, dry contracting band.

Transmission: 18-in. dry-plate clutch through flexible coupling. Selective-type change-speed gearbox with constant-mesh helical gears.

Shipping weight: Approx. 32,500 lb.
Price: £7475

Cable Dozer

General description: Centre-pivoted, angling and tilting blade with triangulated bracing, mounted on "U" frame pivoted to tractor frame, cable-controlled from rear-mounted control unit via overhead conduit.

Blade: Length, 138 ins. (3.5 metres); height, 38½ ins. (.98 metres). Lift above ground, 60 ins. (1.52 metres). Drop below ground, 24 ins. (.61 metres). Blade angle, 25 degrees; tilting adjustment, 7½ ins. (190.5 mm.). *Weight:* 6,200 lb. (2,812 kgs.).

Scraper

Cable-operated. *Capacity* (S.A.E. Standard): struck, 12 cubic yds. (9.17 cubic metres); heaped, 15 cubic yds. (11.46 cubic metres).

Dimensions (overall): Length, 406¾ ins. (10.33 metres); width, 125¼ ins. (3.18 metres); height (blade on ground), 112 ins. (2.84 metres).

General data: Wheelbase, 246½ ins. (6.26 metres). Type of ejection: forward, positive. Apron opening, 70 ins. (1.77 metres). Height of ejector, 70 ins. (1.77 metres). Maximum depth of spread, 12½ ins. (.32 metres). Maximum depth of cut, not limited. Lubrication, pressure grease system. Gauge (centre to centre of tyres): front, 72 ins. (1.83 metres); rear, 78 ins. (1.98 metres). Maximum ground clearance: Rear of bowl, 18 ins. (.45 metres); front axle, 24 ins. (.61 metres). Weight, estimated, 23,400 lb. (10,614 kgs.).

Control Unit

General description: Rear-mounted, spur-gear-driven, double-drum winch.

Line speed (at engine governed speed of 1,800 r.p.m.): Bare drum, 357 ft. per minute (108.8 metres per minute); full drum, 545 ft. per minute (166.1 metres per minute). Clutch, ventilated multi-plate, cam-controlled. Cable size: ½-in. diameter (12.7mm.). *Weight:* 1,700 lb. (771.1 kgs.).

VICKERS VIGOR TRACTOR

Powered by Cummins — **Direct Drive**

The Vickers VIGOR is the fastest heavy track-type tractor in the world; combining great power with high operating speeds, it brings unequalled efficiency to earthmoving operations.

Power operated finger-tip steering with positive featherlight controls for clutches and brakes, provide effortless and accurate manoeuvrability.

With its unique, three-point fully articulated suspension, it enables the reflexing tracks to maintain maximum ground contact area however rough the going. The result is greater traction together with a smoother ride. The constant mesh gearbox permits gear changing on the move, to make full use of the Cummins Diesel Engine's flexible power for tough digging or moving big loads at high speeds up to 9·7 m.p.h.

In design and construction the VIGOR is the world's most advanced heavy track-type tractor; greater operator comfort, better visibility, simplified servicing and maintenance, make it the machine for those who want the best in earthmoving equipment.

VICKERS-ARMSTRONGS (TRACTORS) LIMITED

SCOTSWOOD WORKS, NEWCASTLE UPON TYNE.

Crawler Tractor Scrapbook Pt 2 — Vickers

1. ROLLS-ROYCE ENGINE
2. SUPERCHARGER
3. STARTER MOTOR
4. COOLING SYSTEM
5. AIR CLEANERS
6. PRE-CLEANERS
7. CLUTCH AND CLUTCH STOP
8. GEARBOX
9. BEVEL DRIVE
10. CONTROL BOARD (ELECTRICAL)
11. FUEL TANK
12. SERVO OPERATED STEERING CLUTCHES
13. REDUCTION GEAR AND FINAL DRIVE

Sectional view of the Vickers Vigor Tractor, showing principal units

Vickers VR. 180 equipped with Onions 13 cu. yd. scraper and overhead cable - operated dozer mountings. Other equipment now available for use with the VR. 180 includes other sizes of Onions scrapers, rippers, an extra heavy duty motor grader and vertical boom ditcher. The tractor is powered by a Rolls-Royce supercharged 6 - cylinder Diesel and has a top speed of 9.7 m.p.h.

Farm Mechanization November 1952

"Vickers Vigor" Tractors

VICKERS-ARMSTRONGS (Tractors), Ltd., announce that the Ministry of Supply have placed a further order for 49 "Vickers Vigor" tractors and matched equipment for use by the Royal Engineers.

It was announced in February that an order for 60 had been placed by the Ministry after extensive trials (these have already been delivered to the Army). This new order raises the contract value to the region of £1¼ million.

The "Vickers Vigor" is regarded as the fastest heavy track tractor in the world. It is in use in all six continents on earth-moving, road building and land-clearing projects and is backed by a dealer organisation in more than 50 countries.

Farm Implement & Machinery Review October 1959

The Vickers V-R 180 Tractor

A drawing specially prepared by FARM MECHANIZATION

Manufactured by Vickers-Armstrongs, Ltd., together with a range of earth moving equipment, this tracklayer is powered by a 6-cylinder, supercharged, Rolls-Royce oil engine. Gear changes can be made while the tractor is moving. There are 6 forward and 3 reverse speeds giving a maximum of 9.73 m.p.h. and 7.98 m.p.h. respectively. Steering is by clutch and brake. Fully articulated suspension allows the reflexing tracks to conform to irregular ground. Lubricant is sealed within the track pin bushes. World distribution is by Jack Olding & Co. Ltd., Hatfield, Herts.

CONTROLS

Steering levers, situated in the centre of the driving compartment, operate the steering clutches through the servo unit during the first few inches of their travel, the remaining travel being used to apply the brakes.

Gear levers. The right hand lever is used to select the required gear ratio, the lever on the left controlling the gear range.

A plate indicating the gear lever positions is attached to the front plate of the driving compartment.

CUMMINS DIESEL FOR VICKERS VIGOR

Vickers-Armstrongs (Tractors), Ltd., announce that the Vickers Vigor tractor is now available with a 210-h.p. Cummins diesel engine as an alternative to the Rolls-Royce fitted as standard. Of American origin, the Cummins is now being made in Scotland, and its components are interchangeable with those made in the U.S.A. It is a six-cylinder, turbocharged engine.

Farm Mechanization April 1960

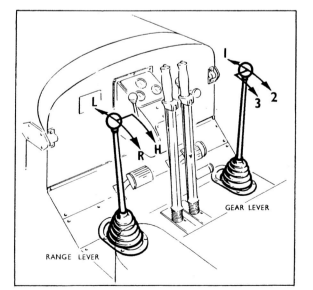

Fig. 2. Gear positions

Crawler Tractor Scrapbook Pt 2 — Vickers

From the family archives of Graham Craw come these two cuttings.
Above: A Vickers Vigor on a road construction site in New Zealand bush country.
Below: Graham's Vickers Vigors line up for a publicity photograph.

FULL SPEED AHEAD WITH THE VICKERS VIGOR

Two Vickers Vigor Tractors operated by Mr G. R. CRAW, of Auckland, preparing the grounds for the Blockhouse Bay School. Mr Craw has recently taken delivery of a third Vickers Vigor.

Reports from big development works all over the world tell the same story. Time schedules on tough jobs are being cut; muck shifting costs are being held down. The Vickers Vigor has proved itself a new force in earth-moving.

A complete range of matched equipment is available to make the most of the tractor's performance. It comes from Vickers-Armstrongs and Onions—famous names in earth-moving. And wherever Vickers Tractors work, skilled service is available with a full range of spare parts.

THE DOMINION MOTORS LTD

VICKERS VIGOR DIRECT DRIVE

SPECIFICATION—3000 SERIES
WITH CUMMINS ENGINE

POWER RATING

Engine horse-power	210
Drawbar horse-power	160

DRAWBAR PULL

The following are calculated drawbar pulls at rated engine speed:

First gear	27600 lb at	2·18 m.p.h.
	12519 kg at	3·5 k.p.h.
Second gear	17500 lb at	3·25 m.p.h.
	7938 kg at	5·23 k.p.h.
Third gear	14400 lb at	3·76 m.p.h.
	6532 kg at	6·04 k.p.h.
Fourth gear	8400 lb at	5·6 m.p.h.
	3810 kg at	9·01 k.p.h.
Fifth gear	6650 lb at	6·53 m.p.h.
	3016 kg at	10·5 k.p.h.
Sixth gear	3250 lb at	9·73 m.p.h.
	1474 kg at	15·65 k.p.h.

At maximum engine torque greater pulling power is developed and the following are calculated values for maximum drawbar pull at reduced travel speeds:

First	31050 lb	14084·0 kg
Second	19716 lb	8942·5 kg
Third	16198 lb	7347·3 kg
Fourth	10195 lb	4624·3 kg
Fifth	7970 lb	3615·1 kg
Sixth	4865 lb	2206·7 kg

TRAVEL SPEEDS AT RATED ENGINE SPEED

	M.P.H.	K.P.H.	F.P.M.	M.P.M.
First	2·18	3·5	192	58·52
Second	3·25	5·23	286	87·16
Third	3·76	6·04	330	100·58
Fourth	5·60	9·01	492	149·96
Fifth	6·53	10·5	579	176·53
Sixth	9·73	15·65	856	260·9
Reverse	2·394	3·843	210	63·91
Reverse	4·122	6·624	363	110·55
Reverse	7·182	11·538	632	192·59

ENGINE

Cummins Model NT-6-BI turbocharged diesel engine with dry sump lubrication. Mechanical variable speed governor set to limit speed at full load to 1,800 r.p.m.

Number of cylinders (vertical in line)	6
Bore	5·125 in. 13·017 cm
Stroke	6·000 in. 15·24 cm
Swept Volume	742·64 cu. in. 12·17 litres

Main Features. Four cycle. Water cooled. Direct injection by Cummins exclusive PT fuel system. Integral cast iron crankcase and cylinder block. Wet cylinder liners. Dry sump lubrication. Oil cooler. Full flow and by-pass oil filters. External viscous silicone type crankshaft damper. Cummins designed and developed turbocharger.

Engine Accessories. Electric starter and dynamo. Hourmeter. Air cleaner, fuel and water filters.

Starting Method	Electric.	Axial engagement.
Starter	5¼ in. (13·33 cm) diameter.	24 volt
Dynamo	5½ in. (13·97 cm) diameter.	24 volt
Batteries	Four 6 volt.	170 ampere hr

DIMENSIONS

Length	14 ft. 4½ in.	438·15 cm
Height (silencer outlet to ground)	9 ft. 3 in.	281·0 cm
Width	8 ft. 8 in.	264·16 cm

Ground Clearance .. 16 in. .. 40·64 cm
(over a width of 25¾ in., 65·4 cm, in centre of tractor)

TRACKS

Width	24 in.	60·96 cm
Gauge (Centre to centre)	6 ft. 8 in.	203·19 cm
Length on ground	8 ft. 8 in.	264·16 cm
Ground pressure	7·11 lb/sq. in.	·499 kg/cm²
Area, ground contact	4992 sq. in.	32206·3 sq. cm
Grouser, height	1¾ in.	4·44 cm
Bolts, diameter	¾ in.	1·91 cm
Pin, diameter	1⅝ in.	4·13 cm
Sleeve, diameter	2¾ in.	6·99 cm

DRAWBAR

Height above ground	17¾ in.	45·09 cm

STEERING

Power Operated	Clutch and brake
Type	Hydraulic
Clutch friction material	Sintered metal
Number of friction surfaces each clutch	22
Brakes	Contracting band

TRANSMISSION

Power transmission, 18 in. (45·72 cm) dry plate clutch, through flexible coupling. Selective-type change-speed gearbox with constant-mesh helical gears.

Gearbox Ratios

First	5·893:1	Sixth	1·323:1
Second	3·961:1	Reverse	5·3:1
Third	3·422:1	Reverse	3·08:1
Fourth	2·3:1	Reverse	1·77:1
Fifth	1·967:1		

CAPACITIES

	Imp. gallons	Litres
Cooling system	10¾	48·9
Crankcase lubricating oil	10	45·46
Gearbox and bevel drive lubriacting oil	4	18·18
Reduction gear and final drive lubricating oil, each	4½	20·46
Fuel tank	75	341·0

WEIGHT—OPERATING

Approximately	35800 lb	16238·6 kg

WEIGHT—SHIPPING

Approximately	35350 lb	16034·48 kg

Vickers-Armstrongs (Tractors) Limited reserve the right to make any changes in the above Specification.

VICKERS VIGOR TORQUE CONVERTER

SPECIFICATION—6000 SERIES
WITH ROLLS-ROYCE ENGINE

POWER RATING

Engine horse-power 210

TRAVEL SPEEDS AT RATED ENGINE SPEED

	M.P.H.	K.P.H.	F.P.M.	M.P.M.
First ..	0—4·66	7·50	0—410	124·97
Second ..	0—6·4	10·30	0—563	171·60
Third ..	0—8·8	14·16	0—774	235·91
Reverse ..	0—5·15	8·29	0—453	138·07

DRAWBAR PULL

Maximum calculated drawbar pulls are as follows:

First gear	42000 lb	19051 kg
Second gear	25000 lb	11339·8 kg
Third gear	16500 lb	7484·3 kg

Maximum pull will depend on traction and weight of fully equipped tractor.

ENGINE

Rolls-Royce Model C6.SFL supercharged diesel engine with dry sump lubrication. All-speed governor set to limit speed under load to 1,800 r.p.m.

Number of cylinders (vertical in line)	.. ..	6
Bore	5·125 in.	13·017 cm
Stroke	6·000 in.	15·24 cm
Swept Volume	742·64 cu. in.	12·17 litres

Main Features. Four cycle. Liquid cooled. Direct injection. Integral cast-iron crankcase and cylinder block. Wet cylinder liners. Dry sump, incorporating oil and water heat-exchanger. Full-flow oil filters. Dipstick, filler tube and cap mounted on sump. External viscous silicone-type crankshaft damper. Power take-off from front end of crankshaft and twin belt pulley for driving the radiator cooling fan.

Engine Accessories. Electric starter. Dynamo. Engine service counter. Positive displacement type super-charger. Triple air cleaners, and fuel filters.

Starting Method	Electric. Axial engagement.
Starter	6 in. (15·24 cm) diameter. 24 volt
Dynamo	5½ in. (13·97 cm) diameter. 24 volt
Batteries	Four 6 volt. 170 ampere hr

DIMENSIONS

Length	14 ft. 4½ in.	438·15 cm
Height	9 ft. 3 in.	281·0 cm
(silencer outlet to ground)		
Width	8 ft. 8 in.	264·16 cm

Ground Clearance .. .. 16 in. 40·64 cm
(over a width of 25¾ in., 65·4 cm, in centre of tractor)

TRACKS

Width	24 in.	60·96 cm
Gauge (centre to centre) ..	6 ft. 8 in.	203·19 cm
Length on ground ..	8 ft. 8 in.	264·16 cm
Ground pressure ..	7·41 lb/sq. in.	·521 kg/cm²
Area, ground contact ..	4992 sq. in.	32206·3 sq. cm
Grouser, height ..	1¾ in.	4·44 cm
Bolts, diameter ..	¾ in.	1·91 cm
Pin, diameter ..	1⅝ in.	4·13 cm
Sleeve, diameter ..	2¾ in.	6·99 cm

DRAWBAR

Height above ground .. 17¾ in. 45·09 cm

STEERING

Power operated	Clutch and brake
Clutch friction material	Sintered metal
Number of friction surfaces each clutch	 22
Type..	Hydraulic
Brakes	Contracting band

TRANSMISSION

Power transmission, 18 in. (45·72 cm) single-plate clutch and hydraulic torque converter through universal coupling. Selective-type three-speed gearbox with constant-mesh helical gears for forward speeds and spur gear for reverse.

Torque Converter

Make	Rolls-Royce
Model	CF 10,000IL
Type..	3-stage
Fluid	Fuel oil
Supply	Charging pump direct from fuel tank through fuel filters

Gearbox Ratios

First	2·22:1
Second	1·643:1
Third	1·177:1
Reverse	2·0:1

CAPACITIES

	Imp. gallons	Litres
Cooling system	13½	61·3
Crankcase lubricating oil ..	7	31·8
Gearbox and bevel drive lubricating oil	4½	20·46
Reduction gear and final drive lubricating oil, each	4½	20·46
Fuel tank	75	341·0
Torque Converter system	5	22·73

WEIGHT—OPERATING

Approximately .. 36000 lb 16329·3 kg

WEIGHT—SHIPPING

Approximately .. 35550 lb 16125·18 kg

Vickers-Armstrongs (Tractors) Limited reserve the right to make any changes in the above Specification.

Crawler Tractor Scrapbook Pt 2 — Vickers

One Vickers Vikon was brought into New Zealand for assessment. It has survived and is owned by a syndicate in Palmerston North, New Zealand. Graham Blackley, who looks after the machine brought it out for photographs in 1998.

Only 21 Vikons were made 16 of which went to overseas destinations. The one that came to New Zealand was a preproduction model made in 1956, serial number 7005 (Vikon serial numbers started at 7000). By December 1957 it was reported to have completed 1198 hours of operation.
Australia received three Vickers Vikons.

The Vickers Vikon came on the market in January, 1958. It was a smaller version of the Vigor and weighed 26,300lbs against 36,000lbs with the Vigor. It was powered by a Rolls-Royce four cylinder engine delivering 142 belt hp and 113 dbhp.

Above: A three quarter rear view of the Vickers Vikon showing the Vickers winch.

Below: The Vikon Rolls-Royce four cylinder engine.

Farm Implement and Machinery Review.—Nov. 1, 1956

Vickers' "Vikon" Tractor

Pre-Production Model Introduced

JACK Olding & Co., Ltd., Hatfield, Herts., exhibited the pre-production model of the new Vickers' "Vikon" tractor for the first time when they showed it at the Public Works & Municipal Services' Congress & Exhibition in London. It is designed and built by Vickers-Armstrong (Engineers), Ltd., and will be supplied with "matched" scraping and dozing equipment.

Powered by a "Rolls-Royce" supercharged 4-cylinder oil engine delivering 142 belt h.p. and 113 drawbar h.p., the "Vikon" has five forward and four reverse gears, with operating speeds up to 8.85 miles per hour and a shipping weight of 26,300lb. Design and construction follow closely on that of the "Vigor," which is the new name for the "VR.180" introduced some six years ago. The 3-point fully-articulated suspension ensures that the tracks maintain their full ground contact area, however rough the "going." The eight large diameter track wheels, including the driving sprocket, have their rims insulated from the hubs by large diameter rubber rings so as to damp excessive vibration. The reflexing tracks have lubricated track pins, the lubricant being retained by rubber seals.

It is claimed that it is this suspension system, proved by hundreds of thousands of hours of actual experience, which enables the operator to employ the machine at high working speeds without discomfort from excessive movement of the hull caused by travelling over rough ground. Moreover, gear changing can be done on the move, for constant mesh helical gears are employed for this purpose. There are two gear levers, one giving four gear ratios, and the other allowing for forward or reverse speeds in each ratio, together with a separate forward fifth gear.

Steering is of the clutch and brake type, with the clutch release hydraulically operated by a servo unit controlled by the steering levers. The initial backward movement of the lever disengages the appropriate steering clutch, while further movement applies the contracting band brakes. The steering system, it is remarked, has the advantage of relieving the operator of unnecessary strain, while affording simple maintenance and servicing.

The operator has a bucket-type seat and an excellent field of vision, a factor which, coupled with the simple controls, reduces operating fatigue and definitely encourages full use of the tractor's capacity right through the working shift.

Unit construction is adopted to give simpler and faster servicing. As an example of this, when the tracks are uncoupled, the tractor can be divided into two principal assemblies, i.e., the power and transmission units, within two hours.

The power-unit is a supercharged 4-cylinder model from the well-known Rolls-Royce "C" range of oil engines, and many of the parts of it are interchangeable with those on the 6-cylinder model powering the "Vigor." An electric starter gives convenient starting up, coupled with operating economy, and power is delivered to the gearbox through an 18in. dry plate clutch and flexible coupling, with the clutch controlled by a foot pedal, which, when depressed, operates a clutch stop to facilitate gear changing on the move.

The "Vikon" Tractor with Earth-Moving Equipment

Crawler Tractor Scrapbook Pt 2 — Vickers

Smaller Vickers Tractor Announced

TRIALS with pre-production models of a new Vickers Tractor have commenced. (Prototype models of the tractor have been in operation for several years.)

The machine is powered by a Rolls-Royce supercharged four-cylinder Diesel engine, delivering 142 belt h.p. Five forward and four reverse gears are provided, giving the tractor operating speeds up to 8.85 m.p.h. The weight is approximately 26,300 lb. A complete range of matched equipment will be produced for this tractor, and will include cable and hydraulic dozers, scrapers, towing winch and logging equipment. It is anticipated that production of the machine will begin in the latter part of 1957. In design, it resembles the larger Vickers tractor, the VR.180, which has been in production for over four years.

Farm Mechanization September 1956

Production versions of this prototype Vickers Vikon are expected to be available in January, 1958. The tractor weighs approximately 26,300 lb. and has a Rolls-Royce, 4-cylinder, Diesel engine which delivers 142 belt h.p.

Farm Mechanization July 1957

Photograph shows a Vikon undergoing trials with an Onions Model 8-11 open-bowl scraper, which will carry a heaped load of 11 cu. yd. The tractor's maximum speed is 8.85 m.p.h.

Farm Mechanization November 1956

June, 1952

More VR-180s

I was interested to hear recently that a smaller edition of the Vickers VR-180 tracklayer is being designed by Vickers-Armstrongs Ltd.

The first production VR-180 was delivered on April 23 with fitting ceremony to Jack Olding and Co., Ltd., Hatfield, who are world distributors for this machine and its earth-moving equipment. When "launching" the tractor, Mr. Henry Hopkinson, who was then Secretary for Overseas Trade, said that it was hoped that an output of 50 VR-180 models would be achieved by October this year. By next year, provided supplies of material and labour were maintained, production would be 1,000.

I am pleased at the determined progress being made with this machine, because it is high time that we challenged our American friends on the business of making giant tracklayers. The VR-180, with its 180 b.h.p. Rolls-Royce oil engine, is not an agricultural tractor in the true sense of the word—although there are large tracts of overseas territories where it could be used as such; but a smaller edition could prove of great interest to agriculture.

Lt.-General Sir Ronald Weeks, chairman of Vickers Ltd., said at Hatfield that the smaller tractor was already in design. He gave no further details, but I think it safe to assume that it will incorporate the articulated suspension and reflexing tracks of the VR-180. An interesting feature of these tracks is that the lubricant is sealed within the track-pin bushes during manufacturing and needs no further attention until a major overhaul is necessary.

These tracks are said to have given a good account of themselves during trials of pre-production models, and it will be interesting to see how they stand up to use by those private owners who, for some inexplicable reason, always seem to impose, without apparently trying, far more strain on a machine than the professional "test-to-destruction driver."

Sir Roland Weeks also mentioned that the trend towards even larger earth-moving machines than the VR-180 was receiving the attention of Vickers-Armstrongs Ltd.

Farm Implement and Machinery Review.—March 1, 1959. 1681

N.I.A.E. Tractor Test Report

"Vickers Vikon" Tracklaying Tractor

(Pre-Production Model)

MADE by Vickers-Armstrongs (Tractors), Ltd., Scotswood Works, Scotswood Road, Newcastle-upon-Tyne, 5. Date of Test: September, 1957. Report No. 164/BS; Test No. BS/NIAE/57/15.

STATIC TILTING TEST

The tractor engine was run at rated speed (without load) for a period of 30min. with the tractor tilted to the maximum angle recommended by the manufacturer for continuous operation as follows:—(a) nose up, 30 deg.; (b) nose down, 30deg.; (c) offside up, 20deg.; (d) offside down, 20deg. During test (a) there was a slight leakage of engine oil (about ¼pt./hour) from the clutch bell housing drain-hole.

GENERAL REMARKS

Repairs and adjustments:—None.
Comments:—None.
Consumption of coolant:—Negligible.
Use of radiator blind/shutters:—Not fitted, thermostat control.
Consumption of engine lubricating oil:—For power take-off engine test, 0.71pt./hour; for 10-hour drawbar test, 0.67pt./hour.
Soil conditions:—The cohesive strength of the soil as used for the drawbar tests was 13.9lb./sq. in. U.S.D.A. classification: clay loam.

BRIEF SPECIFICATION

Fuel:—Diesel oil; specific gravity 0.841 at 60deg. Fah. Cetane No. 54. Nearest U.S. equivalent: commercial diesel fuel.

Tractor:—(Size in accordance with B.S. 2596) Class II, Group 3, Serial No. 7009.

Engine:—Rolls-Royce, Ltd., diesel engine, Serial No. C4-SFL-Proj-2A/13; four cylinders, vertical, in-line; 4-stroke cycle, direct injection; 5⅛in. bore by 6in. stroke, 495.1 cu. in. capacity; compression ratio 14:1; blown aspiration by pressure charger of "Rootes" positive displacement type, manufactured by Sir George Godfrey & Partners, Ltd.; overhead valves; replaceable wet cylinder liners; "C.A.V." mechanical governor, rated engine speed 1,800 revs. per min., governed speed range 450 to 1,800 revs. per min.; "C.A.V." type DFP 3/19 MS. EX. fuel feed pump with one "Purolator" edge-type filter on suction side and two "C.A.V." type 2F 3/4 paper element filters in parallel on pressure side; "C.A.V." type BKB L965684M 4-hole injectors fed by "C.A.V." type NNR4F90/119MGLVWB16M injection pump, serial No. EX197; forced-feed lubrication from gear-type pump with two "Tecalemit" type FG2412 full-flow oil-filters; oil to coolant heat exchanger, S.A.E.30 lubricating oil; pressurised cooling system with centrifugal pump and thermostat for temperature control; "Simms" electric starter motor, excess fuel device for cold starting, four 6v. "Exide" type 3HXAG17KBL lead-acid batteries for starting and lighting, 24v. 136amp./hour system; fuel capacity, 47½gall.; oil capacity, 44pt.; coolant capacity, 6¾gall.; two "Burgess" oil-bath air-cleaners with centrifugal pre-cleaners outside hood.

Transmission and steering:—"Borg & Beck" single-plate 18in. dia. dry clutch, foot-pedal operated; "Vickers" constant mesh gearbox, five forward speeds and four reverse; clutch and brake steering operated by common steering levers, steering clutch actuated by hydraulic servo; brakes manually operated, both steering brakes applied for parking, ratchet on hand levers. Diameters of turning circles: right and left-hand, 19ft. 8in.; rear axle with crown wheel and pinion and spur gear final drive; oil capacities: gearbox, 80 pt., final drives, 36pt. each, S.A.E.90—extreme pressure lubricating oil.

Power take-off arrangements:—Two, one front, one rear, 1.85in. dia. 14-spline giving full engine power, running anti-clockwise viewed from tractor rear, at engine speed.

Sprockets:—Pitch dia. 30.6in., 18 teeth, face width 3.187in.

Tracks:—Non-girder, track pitch 5.312in.; 1.375in. dia. pins; 20in. wide track plates, 54 per track; integral grousers 1.656in. high; track gauge 72.25in., approximate length of track in ground contact 96.5in.

Suspension:—By articulated wheels mounted in pairs; the front pair of wheels on each track are mounted on a cross beam pivoted centrally, the third wheel and driving sprocket are paired and pivot about the reduction gear output shaft. Idler wheel dia. 27in.

Drawbar:—Swinging drawbar, radius of swing 62in., position of pivot centre relative to sprocket centre 37⅜in. forward, lateral adjustment 19½in., no vertical adjustment; height during test 13in. Position of drawbar in accordance with B.S. 2596/1955, A = 1¾in., B = 3½in., C = 3¾in., D = 2in., E = 6in., F = 13in.

Speeds: (at 1,800 revs. per min. rated engine speed):—1st gear 2.02, 2nd 2.95, 3rd 4.23, 4th 5.51, 5th 8.85 miles per hour. Reverse: 1st gear 2.04, 2nd 2.99, 3rd 4.28, 4th 5.59 miles per hour.

Equipment:—No ancillary equipment fitted.

Weight:—Total weight of tractor (including fuel, oil, coolant and operator; weighbridge figures) was 26,020lb.; horizontal distance of point of balance to rear sprocket centre 47⅜in.

Dimensions:—Overall length of tractor, 13ft. 2in.; overall width, 7ft. 8¼in.; overall height, 8ft. 5in. (to top of exhaust); minimum ground clearance, 11in. (to bottom of drawbar).

Vickers Tractor Decision

VISCOUNT Knollys, chairman of Vickers, Ltd., issued the following statement in the company's annual report for the year ended December 31st:—"In view of the heavy expenditure which would be involved in completing a full range of earth-moving equipment in order to meet the highly-competitive conditions now obtaining in that industry, your directors decided to discontinue tractor production when the current programme ended in April, 1961. Losses on termination of tractor production are estimated at £862,000 (£1,750,000 less taxation relief thereon of £888,000) which has been charged to general revenue reserve in the group balance sheet."

The group's Onions Works at Bilston will continue to produce their present range of earth-moving and other equipment under the name of Vickers-Armstrongs (Onions), Ltd.

Farm Implement & Machinery Review June 1961

Crawler Tractor Scrapbook Pt 2

Index

- The machine should always be operated at a speed where it can be correctly controlled. Never do the following:
  - ★ Speeding
  - ★ Sudden starting, sudden braking, sudden turning.
  - ★ Snaking
  - ★ Coasting

- Be careful of those around you, and always confirm that there is no person or obstacle in the way before driving or turning the machine.
- Always operate slowly in crowded places. On haul roads or in narrow places, give way to loaded vehicles.

Allis Chalmers... 6, 10, 11

Bonmartini Count. tracks...16-17,19

Bryden Tracpak...13, 20

Case ...24-31, chart 30-31

Caterpillar...8, 9, 11

Clark...21

Costruzioni Meccaniche Gallamini Vincenzo...18

David Brown...6, 7, 9

Edgeworth Richard L. 1770...14

Farm Mechanization Advisory Service...4, 32, 84

Fiat...33-48, chart 47,48

Fowler... 12, 49-65, 69-73, chart 83

International Harvester...87-104, chart 101,102

John Deere...130-138, chart 131

Komatsu Underwater bulldozer...124-125

Komatsu...10, 115-129, chart 126-128

Komatsu...115-129, chart 126-128

Lanz...139-140, chart 139

Loyd...105-114

Matthews Peter H., tracks...14

Poncet Maurice V., tracks...14

Retreading tracks...15

RNLI...12

Roadless... 24

Rolba Bombardier Muskeg...15

Terratrac...22, 23, 27, 30, 31

Track Marshall...6, 9, 49, 68, 74- chart 83

Vickers Vicaon...155-159

Vickers VR180...141-154, 158, 159

Warco Alwatracs...20

Vietnam

One of our dealers was telling me the other day about the problems in Vietnam preparing ground for forestry. Bomb craters were a big enough hassle for the machinery but where there were no bomb craters things were worse because of unexploded bombs. The Vietnamese were not keen on the craters, they would jump off as the crawler went into the crater and get back on on the other side.

African Bees

African bees do not seem to bother the natives but Europeans dont like them to say the least. The practise adopted when clearing land for forestry, when you knew there were bees in a tree, was to set the crawler in low gear when approaching and jump off before the tree was hit. The driver then walked beside the tractor at a safe distance, until the bees had finished fighting the machine. Sometimes it was necessary to walk considerable distances before the bees cleared off.